How to Use Regents Boosters™

 Flashcards: Complex material is broken down into small chunks of information. The top box on each page is the title; the bottom box contains the info you need to know on the topic. To test yourself, cover the bottom box with the enclosed card cover.

 Hints: Intuitive hints are sprinkled throughout the cards to help you remember the material. There are all different types of hints: acronyms (words made from initials), easy sentences based on the material, and many more.

 Common Regents Question: Alert to common Regents question

 Example: Example follows each concept to show how to apply the material

---

REGENTS BOOSTERS SERIES™

LIVING ENVIRONMENT BOOSTERS™

(Biology)

R. Hertz,
Chief Boosters Pal

Name: _____
(Also known as: Future Living Environment Pro!)

ISBN: 978-0-9747339-7-5

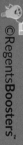 ©RegentsBoosters™

Copyright

Hi there!

I kinda love science and I'm eager to share my excitement and love for science with you.

I teach high school science and discovered that once you explain the material clearly and in an interesting way, the students appreciate and love it too (not to mention that they ace their tests and Regents).

Together with my team of talented Boosters Geeks, we put all that into this fun little book. We hope you enjoy reading it (and laughing through it) as much as we enjoyed creating it!

We love your feedback, so please be in touch!
And welcome to the Boosters Club!

 R. Hertz, Chief Boosters Pal

How to Use Regents Boosters™

 Note: Important detail indicated

 Mr. Boosters: You can depend on him for "peanut gallery" style comments.

 Definitions: The vocabulary word is shown in caps lock, followed by a colon and the definition.

Example — ELEMENTS: The smallest part of matter that can't be broken down into smaller substances

 Hints and funny comments can be found on the right side of the card.

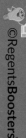

What They're Saying About Regents Boosters™

"I was extremely pleased with the Regents Boosters and highly recommend it. "
—Dr. Gary Schall,
Superintendent of Lawrence Public Schools

"My students loved the hints and easy-to-learn format of the cards. They were able to learn and retain the required material effectively. "
—D. Sprague,
Regents Teacher, Edward R. Murrow High School

"The information is clear cut, without extra info. If you know the book, you know the Regents."
—Brian (Student)

"My son with learning disabilities failed his way through the year. We really wanted him to pass the Regents, but were at loss. We ordered Regents Boosters™ online and... he passed! This is G-d sent."
—Christine (Parent)

These cards are a lifesaver. They are so unique and the hints make it so amazingly simple to remember the material. This is truly a must have for anyone who needs an easier way to retain all the science info in their brains!! Buying this product is...what can I say.. it's a NO BRAINER!!
—Science Geek Not (Student)

COLOR CODED TABLE OF CONTENTS:

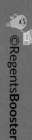

©RegentsBoosters™

One last note before we start:

We will begin learning all the info now.
(I know you can't wait!)

In order to excel on your regents, you must practice many, many regents questions. The more you practice, the better you do.

UNIT 1

CELLS

CELLS

In this chapter we'll master lots of technical info including:

1. Organic compounds

2. Life functions

3. Parts of the cells

4. Enzymes

5. Photosynthesis

6. Cellular Respiration

This unit is very important because it's the basis for a lot of information that we'll learn in future units.

What is a Cell?

A small structure that makes up an organism

(person, animal, or plant).

In other words, all living things are made out of cells.

No, I don't mean a jail cell!

A person's body is made up of brain cells, lung cells, kidney cells, liver cells, muscle cells, skin cells, etc.

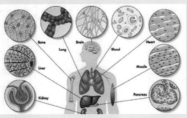

Atom/ Molecule/Compound

A cell is so tiny, you can hardly see it with a naked eye.

But guess what? Cells are made up of even smaller substances!

1. **ATOM:** Smallest chemical element

   • **Types of elements:** carbon, oxygen, hydrogen, nitrogen

2. **MOLECULE:** two or more atoms combined

3. **COMPOUNDS:** two or more elements combined

BONDS: Join atoms together

• Energy is stored in the bonds.

 Each line represents a bond

Great news!
You do not need to memorize the differences or definitions of these 3 terms. You should just recognize these words and know that these are tiny substances that make up cells.

Organic Compounds

ORGANIC COMPOUNDS:
Compound that includes carbon and hydrogen

There are 4 types of organic compounds. Memorize them.
1. <u>S</u>TARCH (made out of simple sugars/glucose)
2. <u>P</u>ROTEIN (made out of amino acids)
3. <u>F</u>AT
4. <u>D</u>NA & RNA

💡 The first letter of each word makes the sentence:
<u>S</u>ome <u>p</u>eople <u>f</u>eel <u>d</u>etermined.

❗ The regents often asks:
1. "Which of the following are organic molecules?"
   The answer would be any of the above four.
2. "Which of the following are <u>in</u>organic molecules?"
   Inorganic molecules mean non-organic molecules (does not contain carbon and hydrogen in the same molecule). Any molecule that is not listed above is inorganic.

Job of Organic Compounds

Organic compounds are attached with bonds.

When these bonds are broken, energy is released for metabolism, building, and fixing.

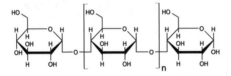

This is a starch molecule. Notice the lines in between the letters. Those are the bonds. When the bonds are broken, energy is released.

Digestion & Synthesis

DIGESTION: Big organic food molecules (think starch, protein, and fat) are broken down in order to enter the cell.

Like this giant sandwich — it needs to be broken down by this guy's teeth to enter his mouth!

SYNTHESIS: Once the broken down molecules (of starch, protein, or fat) enter the cell, the cell will use them as building blocks to build important stuff for our body, like hair and nails.

We need **ENZYMES** to digest and synthesize molecules.

(More on that later.)

Starch Digestion & Synthesis

STARCH DIGESTION: STARCH molecules are too big to enter the cell, so they need to be broken down (DIGESTED) into GLUCOSE (AKA: SIMPLE SUGARS and $C_6H_{12}O_6$)

| STARCH | $\xrightarrow{\text{enzyme}}$ | GLUCOSE | + | GLUCOSE | + | GLUCOSE |

STARCH SYNTHESIS: Once those GLUCOSES get into the cell, they combine together (SYNTHESIS) to create a STARCH

| GLUCOSE | + | GLUCOSE | + | GLUCOSE | $\xrightarrow{\text{enzyme}}$ | STARCH |

Starch/Glucose Names

CARBOHYDRATES refer to all starches and glucose

GLUCOSE has 2 other names:

1. $C_6H_{12}O_6$ (This is the chemical formula)

2. **SIMPLE SUGARS**

The regents sometimes uses these names instead of glucose. So if you see a diagram that has a bunch of $C_6H_{12}O_6$ making up a molecule, you'll know that's a starch.

Examples of starch foods: bread, pasta, cereal

> Nice nickname! Imagine your mom calling you, "Simple Sugar, go study bio!"
>
> PS: If I were you, I'd memorize the nicknames.

Protein Digestion & Synthesis 1

PROTEINS are too big to enter the cell, so they need to be broken down (DIGESTED) into smaller pieces, called AMINO ACIDS.

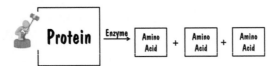

PROTEIN SYNTHESIS: Once those AMINO ACIDS get into the cell, they need to combine together (SYNTHESIS) to create a PROTEIN to do important stuff like building things for our body (like hair and nails)

| | Amino Acid | + | Amino Acid | + | Amino Acid | Enzyme | Protein |

More on Protein

There are many different types of protein because there are many types of amino acids.

So, a protein can be different because it has:

1. Different types of amino acids making it up
2. Different combinations of amino acids
3. Different conformations (shapes of protein) —

If you make different cakes out of different ingredients, the cakes will come out different.

If you make different proteins out of different amino acids (ingredients), the protein will come out different.

⚫ If a protein loses its shape, it won't be able to do its job.

• Examples of protein foods: meat, chicken, eggs, beans

• Proteins are sometimes called ENZYMES.

Fat

- Fats are types of organic molecules.

- They are also called LIPIDS and OILS.

Sorry for bringing up such a sore topic.

- Fats give more energy to the cells than proteins and starches do because they have the most chemical bonds (and energy is released when these bonds break).

Extra Virgin Olive Oil

- Foods containing fats: oil, margarine, butter

DNA & RNA

- These make up our genetic info, which give instructions to the body to do everything it needs to do

- Made up of NUCLEOTIDES

- Located in the NUCLEUS of the cell

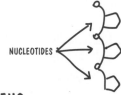

NUCLEOTIDES

NUCLEUS

DNA

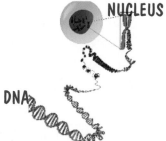

DNA

Life Functions

LIFE FUNCTIONS: In order for something to be considered alive (an organism), it needs to be able to perform these 8 life functions (jobs):

1. NUTRITION/DIGESTION
2. TRANSPORT
3. RESPIRATION
4. SYNTHESIS
5. REGULATION
6. GROWTH
7. REPRODUCTION
8. EXCRETION

🔍 HINT: STRENGRR
— <u>s</u>ynthesis, <u>t</u>ransport, <u>r</u>egulation <u>e</u>xcretion, <u>n</u>utrition, <u>g</u>rowth, <u>r</u>espiration, <u>r</u>eproduction

What are all these stuff? Over the next few cards, we'll explain each one.

A person, animal, or plant can be considered alive because they do all these 8 functions.

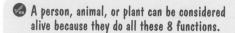

Life Functions (Continued)

LIFE FUNCTIONS can also be called:

1. METABOLIC PROCESSES

2. LIFE PROCESSES

⬤ All life functions interact (work together) in order to maintain homeostasis

Nutrition/Digestion

NUTRITION/DIGESTION: Taking in food and breaking it down

In order for an organism to use the food for energy and building, the food must be broken down first.

- <u>Proteins</u> are too big to enter the cell, so they need to be broken down (DIGESTED) into smaller pieces, called <u>amino acids</u>.

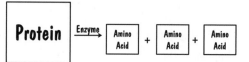

- <u>Starch</u> molecules are too big to enter the cell, so they need to be broken down (DIGESTED) into <u>glucose</u> – $C_6H_{12}O_6$ (AKA: simple sugars)

Transport

TRANSPORT: Materials are transported around the cell

Materials, such as nutrients, gasses, and water are carried all around the cell.

Respiration

RESPIRATION: Producing ATP energy by breaking down nutrients

Our bodies take OXYGEN from breathing and GLUCOSE from eating foods to create ATP energy.

ATP energy is basically energy to do everything we need to do.

Synthesis

SYNTHESIS: Combining smaller molecules to make bigger ones

Think of your little brother building with legos. He is doing synthesis!

Why do we need synthesis (building)?
- For growth (getting taller)
- Grow new cells (hair, nail, and skin cells)

Regulation

REGULATION: Keeping inside conditions normal even when the environment is changing

The inside of an organism (person, animal, or plant) will stay constant even when things are changing outside.

> A person's body temperature is approximately 98.6 °F. When he goes into broiling weather of 104 °F, his body temperature doesn't rise to that number. The body works to maintain its normal temperature of 98.6 °F

We'll be learning tons about homeostasis in the next unit. Stay posted!

Regulation can also be referred to as:
1. HOMEOSTASIS
2. FEEDBACK MECHANISM

Growth

Increase in:

- Size or
- Amount of cells in an organism

Reproduction

REPRODUCTION: Producing new individuals

In other words, organisms having babies

> We'll be learning tons about reproduction in Unit 3. Stay posted!

⬤ In order for a single organism to survive, it must perform all the other life functions. (A cat needs digestion, transport, respiration... to survive.)
BUT if a single organism does not reproduce, it will still survive. (If a single cat does not have babies, it will not die.)

❗ This is a very common regents question.

Excretion

EXCRETION: Getting rid of waste

> ✅ex An organism gets rid of sweat, carbon dioxide, or urine

Life Functions Working Together

All life functions must interact (work together) in order to maintain homeostasis.

Here's some examples that I thought of:

1. Respiration & Excretion: An organism breathes in oxygen with respiration and excretes carbon dioxide as waste.

2. Synthesis & Growth: An organism synthesizes bone cells to help it grow.

Now think of your own examples!

Parts of an Organism

- Organism: a person, animal, or plant
  - ex Mr. Guy
- Organ system: body system
  - ex Digestive System
- Organs: major body parts connected to organ systems
  - ex Stomach
- Tissue: group of similar cells that work together to do a job
  - ex Stomach tissue (a bunch of stomach cells)
- Cell: basic unit of all living things
  - ex Stomach cell
- Organelle: part of the cell

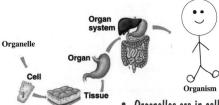

- Organelles are in cells
- Cells are in tissues
- Tissues are in organs
- Organs are in organ systems
- Organ systems are in organisms

Levels of Organization

From smallest to largest:

- <u>In Cells:</u>
    <u>O</u>rganelle ⇨
    <u>C</u>ell ⇨
    <u>T</u>issue ⇨
    <u>O</u>rgan ⇨
    <u>O</u>rgan System ⇨
    <u>O</u>rganism

💡 Acronym: The first letter of each level spells: <u>OCTO$_2$ (pus)</u>

- ◆ There are the most organelles and the least organisms.

- ◆ Organisms are the biggest. Organelles are the smallest.

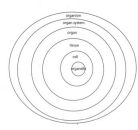

Cells Carry Out Life Functions

We learned all about the different life functions.

Who carries out all these life functions? (Meaning, who is doing respiration, digestion, excretion...?)

Answer: CELLS

🔴 This is a common regents question.

Now let's get a little more nitty gritty.

> Say this 10 times: "Cells carry out life functions."
>
> There's a question on almost every regents asking who carries out life functions.

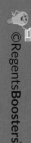

Unicellular Organisms

UNICELLULAR ORGANISMS: Organisms with only ONE cell

MR. UNICELLULAR

ex Algae (type of seaweed) and bacteria

- UNICELLULAR ORGANISMS carry out life functions through their ORGANELLES (parts of the cell).

  (Which makes sense because they only have one cell, so the parts of this one cell will be doing all the hard work.)

This is a picture of a unicellular organism. He is made up of only one cell. The shapes inside are organelles (parts of the cell), which carry out the life functions.

ex Mitochondria does respiration.

Cytoplasm does transport.

Vacuole does digestion.

🔺 There are many regents questions on this topic.

Multicellular Organisms

MULTICELLULAR ORGANISMS: Organisms made up of **MANY** (more than one) cells

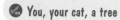 You, your cat, a tree

> 💡 **HINT:** <u>Multi</u> means many. <u>Multicellular</u> organisms have many cells.

- ◆ **MULTICELLULAR ORGANISMS** carry out life functions with their **ORGAN SYSTEMS.**
- ◆ This is a picture of a multicellular organism. He is made up of many cells and organ systems. His <u>organ systems</u> carry out the life functions.

MR. MULTICELLULAR

 Respiratory system does respiration.

Circulatory system does transport.

Digestive system does digestion.

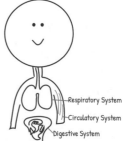

Respiratory System
Circulatory System
Digestive System

🔔 There are many regents questions on this topic.

Similarities Between Unicellulars & Multicellulars

Unicellular and multicellular organisms carry out the **SAME** life functions in **SIMILAR** manners.

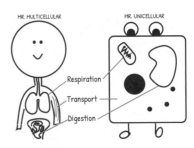

> ✏️ Unicellulars and multicellulars both do respiration. The unicellular organism does it with his mitochondria (organelle) and the multicellular organism does it with his respiratory system (organ system).

Here's a list of some life functions and how unicellulars and multicellulars carry them out:

LIFE FUNCTION	UNICELLULAR ORGANISM	MULTICELLULAR ORGANISM
Transport	Cytoplasm	Circulatory System
Respiration	Mitochondria	Respiratory System
Digestion	Vacuole	Digestive System

❗ There are many regents questions on this topic.

The Cell

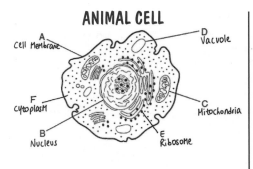

ANIMAL CELL

A Cell Membrane
D Vacuole
F Cytoplasm
C Mitochondria
B Nucleus
E Ribosome

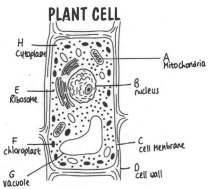

PLANT CELL

H Cytoplasm
A Mitochondria
E Ribosome
B nucleus
F chloroplast
C cell Membrane
G vacuole
D cell wall

The cell has many different parts, called **ORGANELLES.**

🛈 Memorize the pictures of the organelles labeled. Let's explore all these organelles now.

Cell Shape

- Animal cells are <u>round</u>.
- Plant cells are <u>rectangular</u> shaped.

🎤 If you are given a picture on the regents of a blood cell and an onion cell and asked which is which, how will you know?
The blood cell will be the round one (animal cell) and the onion cell will be the rectangular shaped one (plant cell).

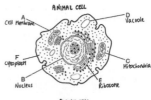

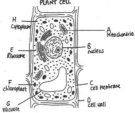

Cytoplasm

CYTOPLASM: A thin gel mostly made up of water with chemicals dissolved in it

- The whole cell is filled up with this thin gel called CYTOPLASM.
- It's job is to transport materials within the cell
- The cell membrane encloses this goo.
- Organelles (part of the cell) float in it.
- You can't see the cytoplasm in the picture because it's represented by the white space in the background of the cell.

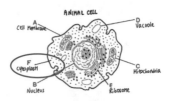

ANIMAL CELL

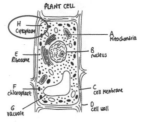

PLANT CELL

Ribosome

- Place where proteins are built
  - The fancy word for that:
    PROTEIN SYNTHESIS: Proteins being built

 OR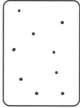

The circles attached to those rod shapes are the ribosomes in an animal cell.

Those tiny circles floating around the plant cells are the ribosomes!

💡 Ribosome-
rib - rib (steak)
is a type of
protein. YUM!

Ribosomes build
proteins.

Mitochondria

• Cell energy (ATP) is created through CELLULAR RESPIRATION in the mitochondria.

 OR

🔑 Mighty mitochondria: Mitochondria produces energy

💡 The mitochondria diagram resembles lightning — a flash of energy. Mitochondria produces energy

Nucleus

- Control center of the cell: gives instructions to other organelles to do different jobs (like protein synthesis)
- Genetic information is stored here in the form of DNA (genes).

 OR

Vacuoles

- Stores materials like waste

- In charge of digesting materials

- Plant cells have <u>large</u> vacuoles.

- Animal cells have <u>small</u> vacuoles.

💡 <u>Vacuoles</u> store waste just like a <u>vacuum cleaner</u> stores waste.

 OR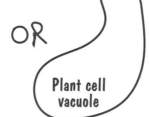

Animal cell vacuole

Plant cell vacuole

Chloroplast

- Changes energy from the <u>sun</u> (solar energy) into <u>food</u> (glucose energy/chemical bond)

- Photosynthesis happens here.

- Contains CHLOROPHYLL

- ONLY in plant cells!

Chloro<u>plast</u>- it lasts. Sun's energy is temporary so it changes it to food which <u>lasts</u> longer.

The filled in ovals are the chloroplast on the diagram.

Cell Wall

Only found in plant cells!
This is the outermost part of the plant cell.

- <u>Jobs:</u>
    1. Protection (against harmful substances)
    2. Structure

💡 A wall provides protection and structure too!

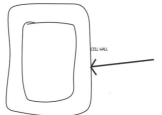

CELL WALL

Cell Membrane

- The layer that surrounds the cell
  - ◆ In an animal cell, it's round
  - ◆ In a plant cell, it's rectangular

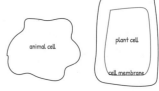

- **SELECTIVELY PERMEABLE:** Lets some things in or out of the cell and not others

- <u>Jobs:</u>
  1. Separates the cell from the environment
  2. Allows movement of materials between the cell and its environment
  3. Helps the cell maintain homeostasis

Cell Membrane Transport

- <u>Job:</u>
  There has to be a perfect amount of "ingredients" inside the cell. Sometimes materials have to come in and sometimes they have to be sent out to maintain homeostasis.

- <u>Done by the processes of:</u>
  1. Diffusion and Osmosis
  2. Active Transport

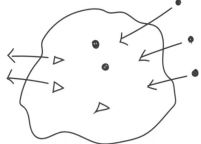

Materials are being transported into the cell and others out of the cell through the cell membrane

Selective Permeability

SELECTIVE PERMEABILITY: Cell membrane selects which molecules can come in and which can come out of the cell.

<u>Questions they ask before they accept a molecule:</u>
* Is it the right Size?
* Concentration: Do I have more in the inside or the outside?

* If there's more of a certain molecule inside the cell, those molecules will come out of the cell.
* If there's more of a certain molecule outside the cell, those molecules will come into the cell.

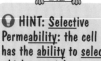

🔵 HINT: <u>Selective Permeability</u>: the cell has the <u>ability</u> to <u>select</u> which materials can enter and which must stay out.

Receptor Molecules

- <u>Where?</u>
  On the surface of the cell membrane

- <u>What Are They?</u>
  Specialized proteins whose shape recognizes
  and attaches to a chemical messengers that
  fit it. Works with shape specificity (see next
  card).

- <u>Job:</u>
  To communicate with the cell to do a job. The
  chemical messenger gives a message to the
  receptor molecule to do a specific job.

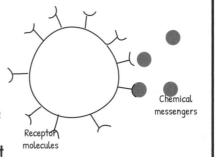

Chemical
messengers

Receptor
molecules

Shape Specificity

SHAPE SPECIFICITY: The protein will only attach onto the chemical messenger if it's an exact match and it has the same shape.

● This is a diagram of an estrogen receptor. Note the specific shape of the receptor site. Only molecules that match that shape perfectly will be able to fit in and cause a reaction.

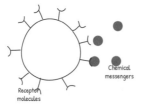

Chemical messengers

Receptor molecules

Who Uses Shape Specificity?

1. Enzymes
2. Antibodies
3. Receptors

• Reactions will only happen when they are matched up with the correct shape.

Note: Although all these substances work with shape specifiity, they each have different functions. Do not confuse them!

❶ This is a common regents question.

Cell Membrane Passive Transport

- Diffusion and osmosis use passive transport.

  - ◆ **DIFFUSION & OSMOSIS:** The process by which molecules move from an area of high concentration to an area of low concentration

💡 No one likes to be in a crowded place.

- **No Energy Is Used In Passive Transport.**

- ❗ This is a common regents question.

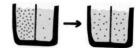

Diffusion & Osmosis

DIFFUSION/OSMOSIS: Movement of molecules from an area of high concentration to an area of low concentration

There are 6 circles inside the cell and 4 circles outside of the cell. Therefore, the circles will diffuse from inside to outside the cell (from high concentration to low concentration)

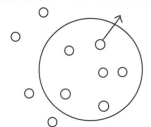

There are 4 circles inside the cell and 6 circles outside of the cell. Therefore, the circles will diffuse from outside to inside the cell (from high concentration to low concentration)

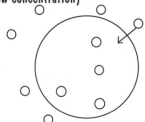

Example of Diffusion/ Osmosis

- If you put any plant in a salt solution, the water moves out of the cell (high concentration of pure water molecules) into the salt solution (low concentration of water molecules). Therefore, a plant that is in a salt solution will shrink, shrivel, or die because the water molecules are moving out.

💡 Salt pulls water to it. Wherever there is more salt, that's where more water will go.

- If you put the cell in distilled water (unsalted water), the distilled water (higher concentration of pure water) will diffuse into the cell (lower concentration of water).

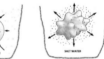

DISTILLED WATER   SALT WATER

- ⚠ This is a common Regents question with the cell of an onion.

Cell Membrane
Active Transport

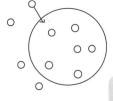

- Molecules move from lower concentration to higher concentration.

- Opposite process of diffusion and osmosis

- Energy of ATP molecules is used

- Usually DIFFUSION & OSMOSIS are used. If the question does not specify which process is used, assume it's diffusion and osmosis. Active transport is only used in specific situations.

Active Transport: This type of transport uses <u>activity</u> (energy).

<u>A</u>ctive <u>t</u>ransport uses <u>ATP</u> energy.

Active Transport for Synthesis

ACTIVE TRANSPORT is used when the cell needs more of a specific molecule (to do synthesis).

Some of the molecules will automatically come in through diffusion, and the rest will be pulled in using active transport.

> If a cell wants to synthesize protein, it will need a lot of Nitrogen. Some Nitrogen will automatically diffuse into the cell, and the extra Nitrogen needed will have to be pulled in using active transport.

Active Transport for Removing Waste

When there is waste in the cell, some will automatically diffuse out of the cell. The extra waste still left will have to be pushed out using active transport.

Catalysts

CATALYSTS: Substances that speed up or slow down chemical reactions without changing themselves.

- Enzymes are a type of catalyst
- Catalysts use SHAPE SPECIFICITY: they "click" into a substance in order to do their job
- Chemical activities need a catalyst to perform all functions
- Reusable: The catalyst doesn't get used up during chemical activity
- Stays the same: The catalyst doesn't change its shape during chemical activity

Enzymes

- Enzymes are <u>proteins</u> with a specific shape that fit a specific substance.

- When the enzyme and substance click together, a chemical reaction happens.

- Once the enzyme did its job and made the reaction happen, it goes to find another substance that fits it to make another reaction happen.

- Jobs: Helps with digestion and synthesis (and many chemical reactions)

- Enzymes cause food to spoil. Therefore, we place the foods in high temperature (cooking) or low temperature (freezing) to slow down or stop the enzymes from causing spoilage.

Enzyme Digestion

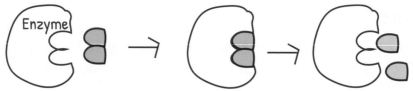

An enzyme and one molecule

The enzyme attaches itself to the molecule

The enzyme splits the molecule into two pieces

Enzyme Synthesis

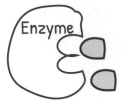  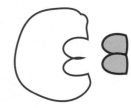

Enzyme

An enzyme and
2 molecules

The enzyme puts
the 2 molecules-
together

Now the two mole-
cules are combined
and the enzyme
looks for another
job

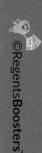

Enzymes Diagram

- ACTIVE SITE: Place on the enzyme where the substrate fits into
- SPECIFICITY: If the shape does not match perfectly, the reaction will not happen.

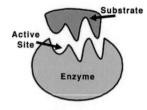

- LOCK AND KEY MODEL: Nickname for the concept of shape specificity

Just like a key matches the keyhole perfectly, the substrate fits exactly into the active site on the enzyme. Therefore, the shape of the enzyme determines whether the reaction can take place or not.

Conditions for Enzymes to Work

1. Correct PH level

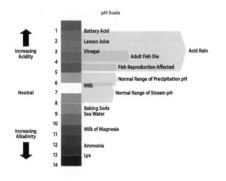

2. Correct Temperature

PH Levels

PH Level: The measurement of how acidic or basic (alkaline) the solution is

<u>PH scale from 0-14</u>
0 = most acidic
14 = most basic
7 = neutral

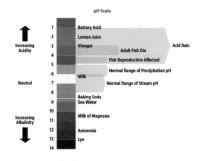

- The PH level affects the enzyme shape.

- Each enzyme functions best (optimally) at a specific PH level. If you increase or decrease that PH level of the enzyme's environment, the enzyme's activity level will slow down or stop.

  ✅ Example: Pepsin is an enzyme found in the stomach that digests protein. Pepsin works best at a PH level of 1.5 to 2 and the stomach's PH level is 1.5 to 3.5. Therefore, pepsin can do its job there. If pepsin travels to the esophagus, which has a PH level of 7, it will not be able to function.

Rate of Enzyme Action

- As the temperature and PH increase, the rate of the enzyme action first increases then decreases.

🛈 This is a common Regents question with a graph.

✏️ Example:

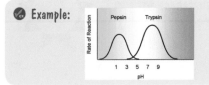

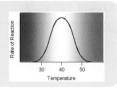

- Each enzyme functions best (optimal level) at a specific PH level and temperature.

- If you increase or decrease the PH level or temperature from the enzyme's optimal levels, the rate of the enzyme reaction will SLOW DOWN or STOP.

- If the temperature gets too hot or the PH level becomes very acidic, the enzyme's shape may change (DENATURE). If that happens, the enzyme will not be able to do its job because it won't fit in to the substance.

Photosynthesis

- The process of the plant changing energy from INORGANIC (oxygen and water) substances to ORGANIC (glucose) substances

-OR-

- The process of using solar energy (energy from the sun) to form chemical bond energies

Photosynthesis

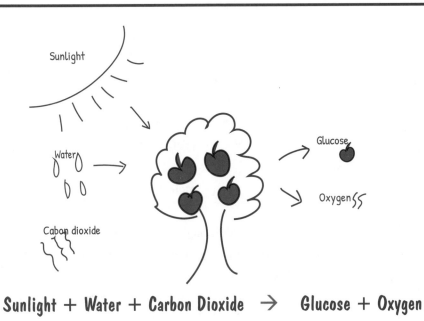

Sunlight + Water + Carbon Dioxide → Glucose + Oxygen

Equation for Photosynthesis

Sun + CO_2 + H_2O $\rightarrow$ $C_6H_{12}O_6$ + O_2

Sun + Carbon Dioxide + Water $\rightarrow$ Glucose + Oxygen

INORGANIC $\rightarrow$ **ORGANIC**
Does not have C and H in the same molecule

Has C and H in the same molecule ($C_6H_{12}O_6$)

- Photosynthesis converts **INORGANIC** materials to **ORGANIC** materials.
- Photosynthesis captures **SOLAR** energy and converts it into **FOOD** energy/**CHEMICAL BOND** energy.
- If you decrease sunlight, carbon dioxide, or water, less glucose and oxygen will be produced.

Photosynthesis
(Continued)

Plants uses energy from the sun to make their food.

 Photosynthesis. When you take a <u>photo</u>, there is a flash of light. Plants use the sun (a source of light) to get the energy to make their food.

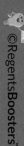

Purpose of Photosynthesis

The purpose of photosynthesis is to make glucose.

<u>For What?</u>
1. Glucose gets stored that the plant can grow.
2. Glucose can be used later in Respiration to make ATP.

Photosynthesis Overview

RAW MATERIALS	WHERE IT TAKES PLACE	SOURCE OF ENERGY	END PRODUCT	WASTE
Carbon Dioxide & Water	In the chloroplast	Solar energy (Therefore, photosynthesis can only happen during daylight hours.)	Glucose	Oxygen (Oxygen and Water get released through stomates as waste)

Photosynthesis Parable

The process of photosynthesis can be compared to cooking a recipe. Follow the chart.

RAW MATERIALS	WHERE IT TAKES PLACE	SOURCE OF ENERGY	END PRODUCT	WASTE
Ingredients	Pot	Fire	Ready made food	Garbage from cooking

Glucose

- Glucose formula

$$C_6 H_{12} O_6$$

- Glucose gets produced through the process of photosynthesis
  -OR-   from digesting a carbohydrate

- Glucose is also called: simple sugar

  - Because plants produce glucose, all plants (potatoes, tomatoes...) produce simple sugars

- Glucose is the #1 source of energy (it stores energy)

- Glucose is food for plants and other organisms

💡 Glucose. Glue. The energy stays glued — it stays stored

- Glucose is called CHEMICAL BOND ENERGY (because energy is stored in its bonds).

- Glucose is FOOD ENERGY because it's energy that's stored in food.

Autotrophic Nutrition

AUTOTROPHIC NUTRITION: Organisms that make their own food

The organisms are called Self-feeding.

- <u>Who?</u>
  Plants and blue green algae

- <u>How?</u>
  It uses the energy of the sun to change inorganic substances (Water and Oxygen) into organic substances (Glucose). (Photosynthesis!)

💡 <u>Auto</u>trophic. They <u>automatically</u> have food because they make it themselves.

Heterotrophic Nutrition

HETEROTROPHIC NUTRITION: Organisms that feed from others because they cannot make their own energy.

- <u>Who?</u>
  Humans and animals

- <u>How?</u>
  They eat other organisms and digest/ break apart bonds to release energy

💡 <u>Heterotrophic</u>. People and animals are referred to as <u>he</u>.

Chloroplast

- Photosynthesis takes place here.

- Located in the leaves

- Contains CHLOROPHYLL: Green pigment in plant cells, which helps change sunlight into food, and gives plant green coloring.

Chlorophyll: (fill) makes up the food to fill up the plant.

Rate of Photosynthesis 1

The Rate of Photosynthesis Depends on:

1. Temperature
2. Amount of light
3. Amount of carbon dioxide & water

💡 Acronym: The first letter of these words spell **TLC** = (**T**emperature, **L**ight, **C**arbon Dioxide and water)

The plants need **TLC** (**T**ender, **l**ove and **c**are) in order for them to be successful with their photosynthesis.

Rate of Photosynthesis 2

- If it is warmer, there is more light, and more carbon dioxide and water available, photosynthesis will happen faster.

- If it is colder, there is less light, and less carbon dioxide and water available, photosynthesis will happen slower.

- If any of these "ingredients" are missing, photosynthesis will stop completely.

Guard Cells

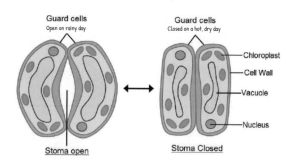

Guard cells
Open on rainy day

Guard cells
Closed on a hot, dry day

—Chloroplast

—Cell Wall

—Vacuole

—Nucleus

Stoma open

Stoma Closed

- The leaves have GUARD CELLS, which change the size of the leaf opening, regulating the exchange of gases and water.

- This is an example of homeostasis: The guard cells open more on rainy days to absorb water and close on hot, dry days to preserve the water. (Very common regents question!!)

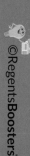

Cellular Respiration

CELLULAR RESPIRATION: The process where cells break up sugars and release ATP energy

- The opposite process of photosynthesis
- Releases ATP energy
- Humans, Animals, <u>and Plants</u> use cellular respiration to get energy.
- Done 24/7 (day and night)
  - ◆ Note: Photosynthesis is only done 12/7, during the day not the night
- Where does cellular respiration happen?
  - ◆ In the MITOCHONDRIA of plants, animals and people
  - ◆ And in the LUNGS of a human or animal
- NOTE: ATP stores AND releases energy.

Equation of Cellular Respiration

$$C_6H_{12}O_6 + O_2 \rightarrow CO_2 + H_2O + \text{ATP Energy}$$

Glucose + Oxygen → Carbon Dioxide + Water + ATP

ORGANIC
Has C and H in the same molecule ($C_6H_{12}O_6$)

→

INORGANIC
Does not have C and H in the same molecule

- Cellular respiration converts **ORGANIC** materials to **INORGANIC** materials.
- Cellular respiration converts **FOOD ENERGY/CHEMICAL BOND** energy to a more usable form of energy (ATP).
- If you decrease the glucose or oxygen, less carbon dioxide, water, and ATP will be produced.

Cellular Respiration

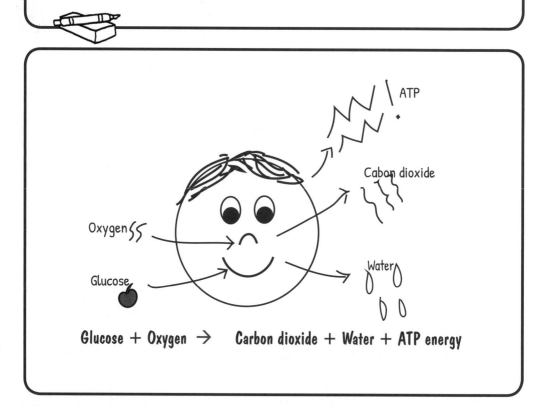

Glucose + Oxygen → Carbon dioxide + Water + ATP energy

Opposite Processes

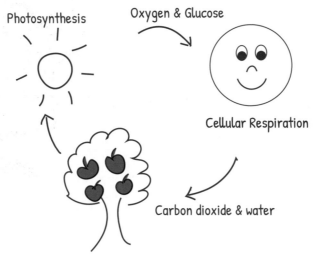

Photosynthesis

Oxygen & Glucose

Cellular Respiration

Carbon dioxide & water

- Photosynthesis gives oxygen and glucose for cellular respiration
- Cellular respiration gives carbon dioxide and water for photosythesis

Congratulations!
You just finished the longest unit!
Keep on going....

UNIT 2

HOMEOSTASIS & BODY SYSTEMS

Homeostasis

HOMEOSTASIS: Organism keeps the conditions inside itself constant EVEN when the environment is changing.

In other words, even when the outside environment is changing, the organism makes sure that everything inside itself stays constant.

⚲ Homeostasis —stay-inside conditions stay the same.

The regents asks many questions on this topic!
They generally ask questions regarding specific examples that we will now explore.

Nicknames of Homeostasis

HOMEOSTASIS has many nicknames and the regents uses these words interchangeably:

1. **DYNAMIC EQUILIBRIUM**
2. **REGULATION**
3. **FEEDBACK MECHANISM** (Body makes a change in response to a change. More on that later.)
4. **FEEDBACK** (same thing as #3)

Now let's explore the examples they ask about homeostasis.

> 💡 <u>Dynamic</u> <u>Equilibrium</u>: <u>Dynamic</u> means changing. Even when the environment is changing, the organism works to keep everything "<u>equal</u>".
>
> 💡 <u>Regulation</u>: Organism is trying to keep everything <u>regular</u>.
>
> 💡 <u>Feedback</u> Mechanism: The body sees a change happening in the environment, so it does an automatic <u>feedback</u> to keep everything normal.

Homeostasis Example #1

Body Temperature

A person's normal body temperature is 98.6°F (37°C).

- On a hot day, when the temperature outside is 101°F, your body doesn't become 101°F. Why not? Your body starts sweating to cool you down to maintain the normal body temperature.
- If it's 32°F outside, you don't freeze. Why? Your body starts shivering and making goosebumps to warm you up and maintain the normal body temperature.

In these examples, your body is doing homeostasis: keeping its temperature constant even though the outside weather changes.

- <u>Change</u>: Temperature
- <u>Response</u>: Sweating or shivering

Homeostasis Example #2

Heart Rate

Imagine you run to school to make it on time to the bell.
You fly in to school and your heart is beating so fast!

Why?
In order to help you run really fast, your body needs more energy. So, your heart starts pumping much faster, which brings more blood all over your body.
This is great because your blood gives energy (oxygen and nutrients) to muscles to move really fast.

So your body responded to a change (homeostasis).
- <u>Change</u>: Running very quickly
- <u>Response</u>: Increase heart rate/pulse rate to bring more energy to your body

Homeostasis Example #3

Breathing Rate

Besides your heart beating, you also get really out of breath from all that running.

Why?

Your body needs more oxygen to run, so you breathe much more intensely. Therefore, you are breathing in more oxygen, which gives your muscles more energy to run. This is great because now your body has more energy to run to school!

By the way, you breathe out more carbon dioxide than usual also because it's the waste product of respiration.

So your body responded to a change (homeostasis).

*   <u>Change</u>: Running very quickly
*   <u>Response</u>: Increase breathing rate brings in more oxygen to the body, which gives the muscles energy to run faster.

Note: Do not confuse! Breathing brings in more oxygen and the heart racing brings more blood which contains oxygen in it.

Homeostasis Example #4

Guard Cells

- The leaves have GUARD CELLS, which change the size of the leaf opening, regulating the exchange of gases and water.

- When it's rainy and cool, the guard cells (and stomates) open to let in the water.

- When it's hot and dry, the guard cells (and stomates) close to keep in the water and not let it dry out.

So the plant responded to a change (homeostasis).
- <u>Change</u>: Rainy or dry
- <u>Response</u>: Guard cells open or close

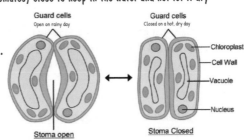

Homeostasis Example #5

Blood Sugar Levels

When you eat a good chocolate bar, your blood sugar level goes up.
This isn't good, because our sugar levels in our blood need to stay normal.

I know school can be stressful!

Our bodies have an organ, called the **PANCREAS** that produces a hormone called **INSULIN**.
Insulin <u>lowers</u> the blood sugar level to keep it normal!

Let's say you skip your breakfast (not a good idea), your blood sugar level gets too low. Our pancreas will release some **GLUCAGON**, a hormone to <u>raise</u> the blood sugar level.

Pancreas

- <u>Change</u>: blood sugar level too high or too low
- <u>Response</u>: Pancreas produces insulin or glucagon.

Blood Sugar Level Diagram

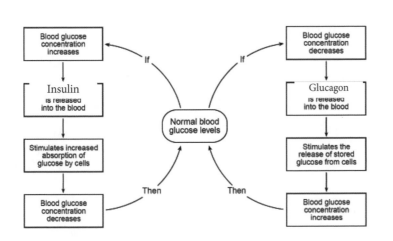

Blood glucose concentration increases

Insulin
is released into the blood

Stimulates increased absorption of glucose by cells

Blood glucose concentration decreases

Normal blood glucose levels

Blood glucose concentration decreases

Glucagon
is released into the blood

Stimulates the release of stored glucose from cells

Blood glucose concentration increases

The regents often asks questions showing the effect of insulin on glucagon and the effect of glucagon on insulin.

Homeostasis Failure

When homeostasis is working, all is fine and dandy.
When homeostasis fails, we get sick.

The regents will ask this in all types of ways. Any time our body is not running the way it should, we know that homeostasis is not working properly.

No, you didn't fail your test. Homeostasis failed.

✅ If the body doesn't sweat, homeostasis failed.

✅ If the body doesn't produce a hormone, homeostasis failed.

There are different things that can cause disease/sickness. We'll explore them over the next few pages.

Cause of Disease: Pathogens

PATHOGEN: Harmful substances that invade the body and cause disease (Think germs)

❶ Pathogens are pathetic gentlemen! They make us get sick!

These pathetic gentlemen (**PATHOGENS**) have some nicknames:

1. MICROBES

2. ANTIGENS
(Sometimes these are bad guys and sometimes they're good guys. When the body recognizes them, they're good. If the body doesn't recognize them, they're bad.)

Pathogens Cause Many Diseases

Types of Diseases Caused By Pathogens:

1. Viruses

✅ Mono, AIDS, and common colds are caused by a virus.

2. Bacteria

✅ Strep throat, staphylococcus aureus, e.coli, salmonella

More on bacteria soon.

3. Fungus/fungi: organism eats off another organism

✅ Athletes foot

4. PARASITE: Organism that harms another organism.

✅ Pinworms

The host is the organism being harmed by the organism.

Note: Viruses, bacteria, and fungus diseases are caused by parasites.

Pathogens & Shape Specificity

In order for a virus and bacteria pathogen to be able to harm an organism, the pathogen needs to be the correct shape. The pathogen shape needs to fit into the receptor molecule in order to harm it.

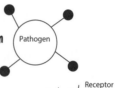

- There are many different pathogens, each with its own shape.
- Each organism has many different receptor molecule shapes as well.
- But, not all organisms have receptor molecules for every pathogen.
- Therefore, there can be a pathogen that will affect a bird, but not a human because the human doesn't have that receptor molecule shape.

Salmonella

Let's talk some more about salmonella, caused by a <u>bacteria</u> pathogen.

Salmonella is a bacteria that comes onto poultry (eggs, turkey, chicken, and meat). When we eat this salmonella, it can make us sick.

- Symptoms: nausea, diarrhea, vomiting, fever
- How to avoid
    1. Cook foods on high temperature for enough time. High temperatures kill bacteria.
    2. Sterilize (wash or soak in hot water) utensils to kill out the bacteria
    3. Do not leave poultry in room temperature. Bacteria grows the most in room temperature.

Antibiotics

Antibiotics kill out <u>bacterial</u> illnesses ONLY. They have saved many lives since they were discovered in the 1900's.

- Problem: ANTIBIOTIC RESISTANCE – bacteria becomes resistant to the antibiotics, which means, the bacteria won't die from the antibiotics.
  Oh no! That means, even if you take the antibiotics, you can still be sick!
- Solution: Take a higher dose or more powerful type of antibiotic
- Problem with that: The bacteria can become resistant to the stronger antibiotic

This cycle can repeat itself many times until we're left with really strong bacteria that we can't kill.

- How can we stop this vicious cycle?
  1. Doctors should not prescribe antibiotics to people who do not have bacterial illnesses (i.e. a cold). The more our bodies get exposed to antibiotics, the bigger the risk will be to develop antibiotic resistance.
  2. Finish your dose of antibiotics! If you only take half your dose, you may feel better, but the stronger bacteria can survive and reproduce. The surviving bacteria can become resistant to antibiotics.

Note: MRSA (Methicillin-Resistant Staphylococcus Aureus) is a form of a Staph infection that is resistant to Methicillin antibiotics. Therefore, it is very difficult to treat.

Cause of Disease: Inheritance

INHERITANCE: If a parent's gene is defective, his child can have this defect too.

- However, if gene isn't defective, the sickness won't be inherited.

> If a man loses his eyesight in an accident, his children can have perfect eyesight because the father's eyesight gene isn't defective.

Cause of Disease: Pollution and Poison

POLLUTION AND POISON: Harmful and poisonous substances

✅ Coal, dust, mercury, lead

🔵 Eating too much tuna isn't good for you since tuna contains mercury. It can cause autism and brain degeneration.

Cause of Disease: Organ Malfunction

ORGAN MALFUNCTIONS: An organ (like the heart, liver, stomach, kidney, lung...) doesn't function properly

 Heart attack

Cause of Disease: Harmful Lifestyles

Harmful Lifestyle		CAUSES:
Smoking	🚬	Lung cancer
Alcohol	🍷	Liver disease
Illegal drugs	💉	Brain damage
Overeating, poor nutrition, and not exercising	🍦	Can cause obesity, which can lead to diabetes, heart problems, high blood pressure

❗ There are many Regents questions on this topic.

So, even if you're nervous for this Regents, slow down on the chocolate!

BMI: Body Mass Index

A Body Mass Index chart is used in order to help doctors and patients figure out if they are underweight, overweight, or just fine.

TABLE 2   Adult BMI Chart

BMI	19	20	21	22	23	24	25	26	27	28	29	30	31	32	33	34	35
Height							Weight in Pounds										
4'10"	91	96	100	105	110	115	119	124	129	134	138	143	148	153	158	162	167
4'11"	94	99	104	109	114	119	124	128	133	138	143	148	153	158	163	168	173
5'	97	102	107	112	118	123	128	133	138	143	148	153	158	163	168	174	179
5'1"	100	106	111	116	122	127	132	137	143	148	153	158	164	169	174	180	185
5'2"	104	109	115	120	126	131	136	142	147	153	158	164	169	175	180	186	191
5'3"	107	113	118	124	130	135	141	146	152	158	163	169	175	180	186	191	197
5'4"	110	116	122	128	134	140	145	151	157	163	169	174	180	186	192	197	204
5'5"	114	120	126	132	138	144	150	156	162	168	174	180	186	192	198	204	210
5'6"	118	124	130	136	142	148	155	161	167	173	179	186	192	198	204	210	216
5'7"	121	127	134	140	146	153	159	166	172	178	185	191	198	204	211	217	223
5'8"	125	131	138	144	151	158	164	171	177	184	190	197	203	210	216	223	230
5'9"	128	135	142	149	155	162	169	176	182	189	196	203	209	216	223	230	236
5'10"	132	139	146	153	160	167	174	181	188	195	202	209	216	222	229	236	243
5'11"	136	143	150	157	165	172	179	186	193	200	208	215	222	229	236	243	250
6'	140	147	154	162	169	177	184	191	199	206	213	221	228	235	242	250	258
6'1"	144	151	159	166	174	182	189	197	204	212	219	227	235	242	250	257	265
6'2"	148	155	163	171	179	186	194	202	210	218	225	233	241	249	256	264	272
6'3"	152	160	168	176	184	192	200	208	216	224	232	240	248	256	264	272	279
			Healthy Weight						Overweight						Obese		

Source: US Department of Health and Human Services, National Institutes of Health, National Health, Lung, and Blood Institute. The Clinical Guidelines on the Identification, Evaluation and Treatment of Overweight and Obesity in Adults: Evidence Report. September 1998 [NIH pub. No. 98-4083].

If a person is obese or very overweight, they have a greater risk of developing diabetes and heart problems. Note: The chart is not always accurate in telling you if someone is overweight or obese since there are other factors, like muscle weight, which can change the results.

How to read the chart:

1. Find your height on the left column. (Sometimes they show the height in inches and you'll have to do some math: 1 foot =12 inches, so 5 feet=60 inches...)

2. Trace your hand across until you reach your weight. (If you are in between two weights given, that's fine)

3. Trace your hand up to find your BMI.

4. Trace your hand down to find which category you are in — healthy, overweight, or obese.

The Immune System

Job: Fight pathogens/fight disease

How?

The immune system has different ways:

1. **WHITE BLOOD CELLS** engulf and kill the antigen
2. White blood cells create an **ANTIBODY**, a shape specific protein that kills the antigen.

We'll explain the first two ways in the next few pages.

3. Inflammation: When you get a cut or a bruise, it can become red/swollen because of the arrival of special white blood cells, which destroy the antigens

4. Fever: the body becomes hot because heat kills the pathogen

White Blood Cells Kill Pathogens
(Immune System)

How does the immune system fight disease? It has two main methods.

Method #1:
WHITE BLOOD CELLS engulf (swallow) the pathogen and destroy it

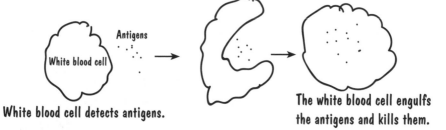

White blood cell detects antigens.

The white blood cell engulfs the antigens and kills them.

These white blood cells are found in the blood and play a major role in protecting our bodies.

Antibodies Kill Pathogens
(Immune System)

Method #2 to kill pathogens:

White blood cells create **ANTIBODIES**, shape specific proteins to fight each specific antigen (pathogen). These antibodies are like custom-made weapons to kill out specific enemies.

- Every antigen (i.e. a cold, chicken pox) has a specific shape.
- Therefore, each time an antigen comes to visit our body, the white blood cells create a custom-made weapon, called an antibody, to kill out that specific antigen.
- If that antigen ever revisits, the body will pull out the antibodies it made a while back and use them against the antigen! And you won't get sick from that specific antigen again!

Ⓠ <u>Anti</u> -<u>body</u>. These substances are <u>anti</u> the bad guys and try to protect the <u>body</u>.

Antigens

Antibodies

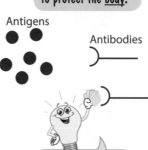

Do not confuse antigen and antibody!

Antibody Shape Specificity
(Immune System)

Each antibody will only work for one type of antigen. If a different antigen visits, it won't work. It will need to create new special antibody for him.

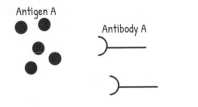

If antigen A visits our body, the body will create antibody A, a custom-made weapon to fit the shape of antigen A.

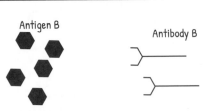

If antigen B visits our body, antibody A will **NOT** be able to fight it because it's the wrong shape! Therefore, the body creates a new antibody, antibody B, that matches the shape of this new enemy.

Antigen
(Immune System)

ANTIGEN: A protein marker on the outside of the cell that checks if the cell belongs to the body, or doesn't belong.

- If the antigen matches the body's marker, the immune system knows not to attack it.

- If the antigen doesn't match the body's marker, the immune system will view it as a pathogen and will try to attack it.

    Foreign antigens: outside antigens that the body does not recognize

- Note: The regents sometimes refers to antigens as pathogens.

Vaccine
(Immune System)

1. A vaccine contains a dead or weak strain of a specific pathogen (antigen) and is injected into the body. You don't get the sickness because it's a dead or weak strain.

   🖊 Chicken pox vaccine contains dead or weak chicken pox antigens.

2. The body's immune response: produces antibodies to fight these antigens

   🖊 The body produces antibodies for chicken pox.

3. If this <u>specific</u> type of sickness will try to attack your body in the future, your body will have the antibodies to fight them!

   🖊 You will be immune to chicken pox, but not the mumps, from this vaccine.

⬤ Each new time you get a new cold, it's a different strain of the cold virus since your body already created antibodies to fight the other colds that you had in the past.

⬤ Each year a new strain of the flu comes out. Therefore, we need to get a new flu shot each year.

⬤ There are many Regents questions on this topic.

Problems in the Immune System 1

- **ALLERGIC REACTIONS:**
  The body thinks that "innocent guys" (HARMLESS ANTIGENS) like mold, dust, and certain foods (peanut butter) are "bad guys", so the body fights them. The body creates:
  1. Antibodies against them and tries to kill them &
  2. Histamines which causes sneezing, difficulty breathing, and rashes.
     - This is an example of homeostasis failing/immune system not working properly.

- **AUTOIMMUNE DISEASE:**
  The body rejects its own cells because it doesn't recognize its own antigens anymore. Then, the body attacks itself.

Problems in the Immune System 2

AIDS/HIV:

- ◆ Attacks white blood cells → Immune system stops working → Body can't fight sicknesses like pneumonia and cancer.
- ◆ Caused by the HIV virus
- ◆ Contagious through blood and body fluids
- ◆ Prevention: Wear gloves when dealing with blood, doctors should use new needles for each person when giving shots, never use drugs and don't share needles

Transplants
(Immune System)

Transplants: If a person's organ fails, they can get a donated organ from a healthy person.

- **Problem:**
  When you transplant any organ (kidney, lung, liver...), the immune system can reject the organ because of its unfamiliar antigens.

- **Solution:**
  Take medicine (IMMUNOSUPPRESSANTS) that weakens the immune system so that it won't reject this organ.

- **Danger:**
  Immune system won't be strong enough to fight other sicknesses. People who take immunosuppressants can become fatally ill from pneumonia or even a bad cold.

Digestive System

DIGESTION: Breaking down the food into small pieces so that the cell can absorb it.

- The body cannot digest big pieces of food, so the body needs to break it down into smaller pieces that it can enter the cells in the body.
- The digestive system includes: mouth, esophagus, stomach, small intestine, large intestine.
- Processes that happen when you eat:
  1. **DIGESTION:** food gets broken down
  2. **ABSORPTION:** nutrients get absorbed into your blood
  3. **CIRCULATION:** the blood brings the food all over the body
  4. **DIFFUSION:** some nutrients enter the cells, and others are sent out of cells

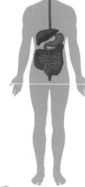

Memorize DACD: Digestion, Absorption, Circulation, Diffusion

Main Nutrients
(Digestive System)

Our bodies need specific types of foods in order to operate properly:

1. **PROTEIN** helps our bodies build things (hair, nails, growth)
2. **CARBS** give us energy
3. **FRUITS** and **VEGGIES** give us nutrients

Problems in the Digestive System

- **ULCERS:** Gastric juices burn through the lining in the stomach
- **COLON CANCERS:** Stool moves too slowly and causes cancer in the large intestine.
- **CONSTIPATION:** The stool is too dry and thick to pass through the excretory system, so it gets stuck.

🔵 It is very important to get enough **FIBER** and **ROUGHAGE** (food that doesn't get digested, loosens the stool, and helps waste be released) in your diet to prevent constipation.

> ✅ Apples, pears, beans, whole grains, bran

Respiratory System

<u>Job:</u> **GAS EXCHANGE: We breathe in oxygen and breathe out carbon dioxide**

- Organs used: Nose, lungs, diaphragm (muscle under the ribcage that moves the air in and out of lungs)
- Cellular respiration, which produces ATP energy, occurs in the lungs (in multicellular organisms)

<u>Vocabulary Time!</u>
- **INHALE:** breathe in (oxygen)
- **EXHALE:** breathe out (carbon dioxide)

Respiratory System

Cellular Respiration
(Respiratory System)

- Breathing is a part of the process of cellular respiration, which creates ATP energy.
- Cellular respiration happens in the LUNGS (in multicellular organisms)

Glucose + Oxygen → Carbon dioxide + Water + ATP energy

Problems in Respiratory System

<u>Sicknesses:</u>
Bronchitis, Asthma, Pneumonia, lung disease,
Low lung capacity: amount of air the lungs can hold

<u>Main Cause:</u>
Smoking

Circulatory System

- **JOB:** Transports materials (nutrients, gasses: oxygen & carbon dioxide, water, glucose) all over your body through the **BLOOD**

- The **HEART** pumps the blood

- **VEINS** and **ARTERIES** are the "pipes" where the blood flows through.

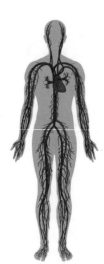

Circulatory System

Pulse Rate
(Circulatory System)

- After the heart pumps the blood, it travels through your whole body.
- You can feel your blood pulsating in your wrist, neck, and other places.
- If you count the amount of pumps per minute, this will give you your pulse rate.
- You can also count your pulse rate for 20 seconds and then multiply that number by 3. This will give you your pulse per minute.

> You count 30 pulses in 20 seconds.
> 30 x 3 = 90. Your pulse rate is 90 per minute

- If you increase your activity, your heart rate goes up, and so does your pulse rate

Parts of Blood
(Circulatory System)

1. **Plasma:** liquid part of blood, transports materials
2. **White Blood Cells:** fights diseases
3. **Red Blood Cells:** carries oxygen
4. **Platelets:** help the blood clot

White blood cells fight disease. White represents purity and cleanliness.

 Red Blood cells: transport oxygen

 White Blood cells: fight pathogens

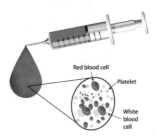

Red blood cell
Platelet
White blood cell

Problems in Circulatory System

- Clogged arteries: there's a blockage in the "pipes", so the blood cannot flow through
  - Leads to: Heart attack, high blood pressure, and stroke

- Heart attack
  - Cause: Overweight, smoking

- High blood pressure

- STROKE: arteries to the brain are blocked so brain doesn't get oxygen

- Blood Doping: Athletes give themselves red blood cell transfusions. Red blood cells contain oxygen, which gives energy to the muscles. This is illegal and very dangerous because it can make the blood too thick and cause blood clotting.

Excretory System

Waste is removed from body

Organ in Excretory System	What it excretes (Excretory waste)
Kidney	Urine
Lung	Carbon dioxide + Water
Skin	Sweat

BLADDER: Urine is stored here until it leaves the body.

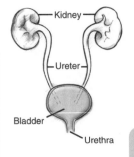

Kidney
Ureter
Bladder
Urethra

💡 Excretory - <u>X</u>: All the things that are <u>X</u> needed anymore.

Kidney

- Removes waste (like urea in the form of urine)

- Regulates salt and water levels

- Chooses what's waste and what's needed

- The kidney has similar function to the cell membrane, which filters what comes in and out of the cell.

Homeostasis in the Excretory System

Homeostasis in the excretory system: Regulates the amount of each substance you have in your body

✅ The more water you drink, the more urine output there is.

✅ When you sweat, you lose water → the endocrine and nervous system send a hormone to the kidney to return water from urine into blood, which makes the urine more concentrated and dark.

Homeostasis in the Skin
(Excretory System)

Homeostasis of the Skin

Sweating
1. Removes nitrogen waste
2. Regulates body temperatures

Problems in Excretory System

- **KIDNEY STONES:** Too much calcium in the kidneys
  Causes: Eating too much protein, drinking too little water

- **NEPHRITIS:** Inflammation in the kidney

⛔ There are hardly any regents questions about the Excretory System problems.

Nervous System

<u>Jobs:</u> Receives and sends impulses (messages to the body to do things)

> ⚡ Your hand touches a hot pot and sends a message to your brain that the pot is hot.
>
> Your brain then sends a message to your hand to move it away from the pot.

- **CELLULAR COMMUNICATION:** Nerve cells communicate with each other to carry out different jobs around the body
- The brain detects a **STIMULI** (change) and sends a message to the cells to react
- The nervous system controls most actions in the body, including movement, balance, and reactions to changes in the environment

Parts of the Nervous System

1. **BRAIN:** receives the messages and sends messages to the body to do different things

2. **SPINAL CORD:** long, fragile tube-like structure that begins at the end of the brain stem and continues down almost to the bottom of the spine
   - The messages are sent from the brain down the spinal cord and then to the nerve cells.
   - If the spinal cord is damaged, the messages cannot be sent down and the injured person can lose control over the parts past the injury. (This is an example of homeostasis failure)

3. **NEURONS:** nerve cells — messages are sent to nerve cells found all over our body

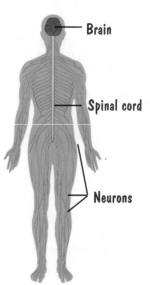

Brain

Spinal cord

Neurons

How the Nervous System Sends Messages: Method 1

Direct contact between nerve cells: The brain sends message to a body part to do something. How is the message going to reach that body part? We have some great messengers called nerve cells. Each nerve cell touches the next nerve cell, and while it's touching the next nerve cell, it sends the message until it reaches the body part that needs to do the action.

ex Let's say the brain needs to tell the hand to move away from the hot pot. The brain sends a message to nerve cell A which delivers the message to nerve cell B, which delivers the message to nerve cell C, which delivers the message to nerve cell D... until it reaches the hand and gives it the message to move away from the hot pot.

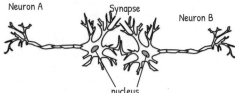

Neuron A

Synapse

Neuron B

nucleus

The place where the nerve cells meet is called the **SYNAPSE**.

In this picture, each person is delivering a message to the next person. Neurons send messages to the next neuron in a similar way!

How the Nervous System Sends Messages: Method 2

The brain sends a messenger to a nerve cell to tell him to do something.
- The messengers are called **CHEMICAL MESSENGERS** (molecules that send messages to a nerve cell to do something)
  Note: Chemical messengers are often **HORMONES**
- The **RECEPTOR MOLECULE** (found on the outside of the cell) accepts the chemical messengers that match its shape and then delivers the message to the cell
- **SYNAPSE:** Place where chemical messenger and receptor molecule meet and send the message

- Shape specificity: A chemical messenger will only be able to communicate with a cell if it has a receptor molecule that matches its shape.
- If there are not enough receptor molecules, the chemical messengers will not be able to carry out it's message.
- If there are not enough chemical messengers, the receptor molecules will not be able to carry out the message.

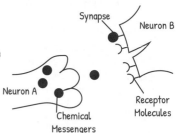

Neuron A produces chemical messengers. Neuron B has receptor molecules that match that chemical messenger.

Stimulus / Response

- **STIMULUS: A factor that causes the organism to react**

  ✅ You touch a hot pot

  The hot pot is a STIMULUS.

- **RESPONSE: How the organism reacts**

  ✅ Pulling your hand away is the response.

Problems in the Nervous System

1. The spinal cord is injured, so the messages cannot be sent from the brain down the body.
2. Nerve damage will cause a person not to feel things and get messages to respond to changes.
3. Brain damage can cause a person to lose body functions because the brain can lose its ability to send messages to the body to do different things

Endocrine System

Job: Secretes (produces) **HORMONES** for cellular communication to regulate the body

- Controlled by **FEEDBACK MECHANISM**
- Works hand in hand with the nervous system
- Hormones can only do their job if they attach to a receptor molecule with the correct shape (shape specificity)
- **GLANDS** are organs that produce hormones

Vocabulary time: secretes a hormone/inhibits a hormone
- Secretes: produces (makes more of)
- Inhibits: prevents (makes less of)

Glands in the Endocrine System

Here's a list of glands and which hormone each gland produces:

ENDOCRINE SYSTEM

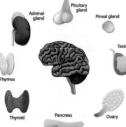

GLAND	HORMONE
Pancreas	Insulin and glucagon
Ovaries	Estrogen & Progesterone (females)
Testes	Testosterone (males)
Pituitary Gland	Controls other glands

PITUITARY GLAND

This diagram shows how the pituitary gland controls other glands. You don't need to memorize the glands. Just understand the concept.

Blood Sugar Levels
(Endocrine System)

A normal amount of glucose in the blood means normal blood sugar level.

- When there's a lot of glucose (from sugary foods), insulin (a hormone) comes and prompts glucose to move out of the blood into the liver → lowers glucose in blood until it reaches a normal level (homeostasis).

- When the glucose level becomes too low, glucagon (a hormone) prompts the release of glucose (stored in the liver) into the blood → raises glucose in blood until it reaches a normal level (homeostasis).

Problems in the Endocrine System

1. **DIABETES:**
- The **PANCREAS** is supposed to produce **INSULIN**, which lowers blood sugar level to make it a normal level.
- A diabetic person's pancreas doesn't produce insulin (or the body doesn't respond to the insulin) → the person has high blood sugar levels. This can be controlled by a special diet or insulin injections.

2. Not enough growth hormone produced. A growth hormone shot can be administered.

Muscular & Skeletal Systems

MUSCULAR SYSTEM: Controls the muscles in the body

> 🔵 ex. Walking, running, writing

SKELETAL SYSTEM: Controls the bones in the body

> 🔵 ex. Standing, walking, running

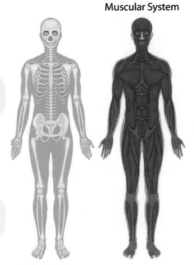

Muscular System

Skeletal System

Homeostasis & Body Systems

All body systems work together in order to maintain homeostasis.

Here are some examples:

1. Digestive system & Excretory System: the digestive system digests the food and then sends the waste out through the excretory system.

2. Respiratory System & Circulatory System: the respiratory system regulates the inhaling of oxygen and exhaling carbon dioxide, while the circulatory system transports these gasses around the body.

3. Nervous System & Endocrine System: The nervous system sends messages to glands in the body to produce hormones.

4. Muscular & Skeletal System: The muscular system regulates the muscles which can move the bones in the skeletal system.

Life Functions & Body Systems

Unicellular and multicellular organisms carry out the SAME life functions in SIMILAR manners.

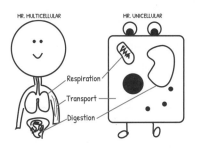

MR. MULTICELLULAR   MR. UNICELLULAR

Respiration
Transport
Digestion

Here's a list of some life functions and how unicellulars and multicellulars carry them out:

LIFE FUNCTION	UNICELLULAR ORGANISM	MULTICELLULAR ORGANISM	ORGAN
Transport	Cytoplasm	Circulatory System	Blood
Respiration	Mitochondria	Respiratory System	Lungs
Digestion	Vacuole	Digestive System	Stomach
Excretion	Cell membrane	Excretory System	Kidney

UNIT 3

REPRODUCTION

Reproduction

REPRODUCTION: Producing offspring or new individuals through the sexual or asexual process

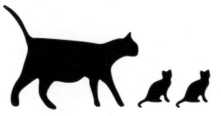

Building Materials: DNA

Whenever something new is produced (in reproduction), the DNA (genes) get copied.

- The **CHROMOSOMES** are sticks of DNA

- Each organism (living thing) has its own normal amount of chromosomes (which are found in the nucleus)

💡 Chromo<u>some</u>, There are <u>some</u> DNA in every chromosome.

> ✔ex Humans have 46 chromosomes, frogs have 26 chromosomes, and lions have 38 chromosomes

DNA

Asexual Reproduction

ASEXUAL REPRODUCTION: One cell divides into two identical cells.

- Uses the process of **MITOSIS/CHROMOSOMAL REPLICATION:** The cells make an identical copy of themselves

- **No GENETIC VARIABILITY:** the genes of the original cell are 100% identical to the new cells. Therefore, there's no differences in the genes = no genetic variability.

- **PARENT CELLS** produce **DAUGHTER CELLS**

💡 Mitosis. <u>Toes</u> —You have <u>two</u> sets of <u>toes</u>. By mi<u>tos</u>is, the cell changes from one set of chromosomes to <u>two</u> sets.

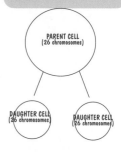

PARENT CELL
(26 chromosomes)

DAUGHTER CELL
(26 chromosomes)

DAUGHTER CELL
(26 chromosomes)

Parent & Daughter Cells

PARENT CELL: The original cell
DAUGHTER CELLS: The two new cells produced

- Daughter cells are also known as **CLONES** because the daughter cells are genetically exactly the same as the parent cell.

- Daughter cells have the same number of chromosomes and the same characteristics as parent cells.

- Produced by asexual reproduction

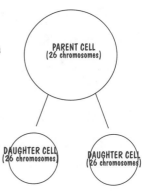

PARENT CELL
(26 chromosomes)

DAUGHTER CELL
(26 chromosomes)

DAUGHTER CELL
(26 chromosomes)

Types of Asexual Reproduction

1. BINARY FISSION: Cell divides into two pieces.

✅ Bacteria: one cell divides into two cells

2. VEGETATIVE PROPAGATION: A new plant grows from a piece of the original plant

✅ If you cut a branch off of a tree and replant it, the branch will grow identical to the original tree!

3. BUDDING: Organism forms a growth and then breaks off

✅ Yeast: The yeast starts off as an oval shape, then forms a growth. The growth breaks off and becomes a new yeast cell.

4. Organism splits into pieces and develops as 2 separate organisms

✅ If you cut a worm in half, both parts of the worm will be alive! You now have 2 worms!

Asexual Reproduction Examples

1. Mold

2. Nail and hair growth

3. Zygote formation: after fertilization, the fetus' cells reproduce with asexual reproduction/mitosis

   (More on that later)

4. Cancer

Yup! That's how that mold grows all over your bread. The mold keeps reproducing and reproducing asexually.

Cancer: Asexual Reproduction

CANCER: Uncontrolled cell division in a specific tissue

> ✅ Skin cancer=uncontrolled cell division (too many abnormal cells being produced) in the skin tissue

- One cancer (abnormal) cell keeps reproducing and making more abnormal cancer cells
- If the cancer is in a **BODY** cell, it will not be hereditary.
- Leukemia damages the immune system.
- Treatments:
    1. Chemotherapy: stops all mitotic cell division to make sure the cancer stops growing. But it also stops normal mitosis, and that's why chemo patients can lose their hair.
    2. Radiation: kills cancer cells directly with radiation
    3. Surgery: manually removes the cancer. But, they need to be sure that every cancer cell is removed. If even one cancer cell remains, it can reproduce and reproduce...
- **CARCINOGEN:** things that cause cancer

> ✅ Smoking causes lung cancer. Tanning can cause skin cancer.

⬤ A healthy immune system will fight cancerous cells before they start to reproduce and become dangerous.

Mitosis Steps

1. **CHROMOSOMAL REPLICATION** (the chromosomes/DNA make an exact copy of themselves)
2. Chromosomes line up in a single file
3. Strands separate and move apart
4. Final result: 4 single strands of chromosomes with the same exact genetic information

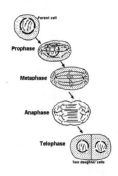

Notes:

- Mitosis is sometimes called **MITOTIC CELL DIVISION**
- Since asexual reproduction is done through mitosis, the regents uses the words mitosis and asexual reproduction interchangeably
- You do not need to know the names of the steps (prophase, metaphase...)
- The regents may ask you to draw the daughter cell. You should draw a cell identical to the original cell.

Cloning
(Asexual Reproduction)

Scientists can now CLONE — "make a copy" of an animal using asexual reproduction (mitosis).

Let's learn how they do this.

PS, the regents often asks about cloning animals in relation to sheep, since they were successful in cloning sheep.

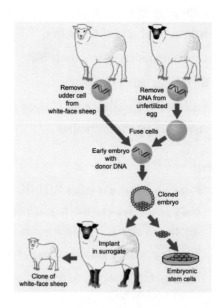

Remove udder cell from white-face sheep

Remove DNA from unfertilized egg

Fuse cells

Early embryo with donor DNA

Cloned embryo

Implant in surrogate

Clone of white-face sheep

Embryonic stem cells

Cloning Process
(Asexual Reproduction)

🔵 Please follow along with the diagram on the previous page.

1. Nucleus of a <u>body</u> (somatic) cell is removed from organism #1.

   This is the body you want to clone. You now have the genetic info of this body.

2. The nucleus and genetic material of an <u>unfertilized egg</u> is removed from an organism #2.

3. Fuse the body cell from organism #1 and the "empty" egg cell of organism #2 using electricity.

4. Insert this combination of a body cell and an unfertilized egg into a uterus of organism #3.

<u>Result</u>: This new creation will be identical to organism #1 where you took the body cell from (and not to organisms #2 and #3).

🔴 Be careful— often cloning just means mitosis. They only mean this type of cloning sometimes.

Sexual Reproduction

Sexual Reproduction: Two organisms produce a new organism (a new baby!)

In order for a new baby to form, these 5 processes need to happen:

1. MEIOSIS
2. FERTILIZATION
3. MITOSIS
4. DIFFERENTIATION
5. GROWTH

What are all these stuff? Let's learn!

Who uses Sexual Reproduction?

Multicellular organisms when forming offspring

• Humans, Animals, and plants (most of the time)

Important:

• With sexual reproduction, there is GENETIC VARIABILITY: varied (different) genetic characteristics in the population because the kid inherits half the genes from the mother and half from the father.

• Asexual reproduction has no genetic variability.

Meiosis
(Sexual Reproduction)

MEIOSIS: A body cell splits to form a GAMETE / SEX CELL
- This process occurs in the male and female bodies individually
    1. When a <u>male</u> body cell splits, it becomes a SPERM cell
    2. When a <u>female</u> body cell splits, it becomes an EGG cell
- Sperm cells and egg cells are types of gametes/sex cells

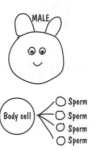

Meiosis
(Sexual Reproduction)

In meiosis, one body cell splits into <u>four</u>.
- In a male, one body cell becomes 4 healthy sperm cells
- In a female, one body cell becomes 1 healthy egg cell and 3 egg cells that are not usable.

The gamete/sex cell/sperm/egg cell contains HALF the number of chromosomes than a body cell.
So, since a human body cell contains 46 chromosomes, a human gamete contains 23 chromosomes.

HAPLOID/MONOPLOID: Half the chromosomes of the body cell
- the haploid = the amount of chromosomes found in a gamete/sex cell/sperm/egg cell
- 'n' represents the haploid number

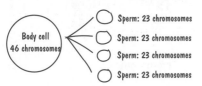

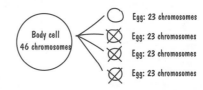

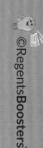

Meiosis Process
(Sexual Reproduction)

1. Chromosomal/DNA replication: The DNA makes a copy of itself
   (Example: 46 chromosomes x 2 = 92 chromosomes)
   The chromosomes cross over here (Recombination).
2. The cell divides once
   (Example: 92 chromosomes / 2 = 46 chromosomes)
3. Cell divides again
   (Example: 46 chromosomes / 2 = 23 chromosomes)

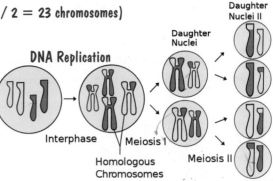

DNA Replication

Interphase   Meiosis I

Daughter Nuclei

Daughter Nuclei II

Meiosis II

Homologous Chromosomes

Meiosis Process
(Sexual Reproduction)

Notes:

- DNA/chromosomal replication happens in both meiosis and mitosis

- Meiosis is sometimes called **MEIOTIC CELL DIVISION**

- The regents may show you **DNA** in the original cell and ask you to draw how the cell will look after meiosis. You will draw **HALF** the shapes of the original cell.

- You do not need to memorize the meiosis step names (prophase, metaphase...)

Recombination/Crossing Over
(Sexual Reproduction)

RECOMBINATION: During DNA replication in meiosis, the chromosomes cross over and form a whole new set of genes.

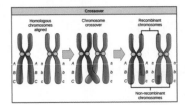

This is how we get so many different gene combinations in organisms that reproduce with sexual reproduction.

This is why a person can have children who have totally different characteristics than their mother and father. The genes of both the father and mother crossed over (separately) during meiosis.

Fertilization
(Sexual Reproduction)

FERTILIZATION: The sperm from the male and egg from the female unite to form a zygote (the beginning of a baby).

Sperm cell   +   Egg cell $\longrightarrow$ Zygote

Note: In this diagram, the egg cell and zygote look the same, but they are totally different!

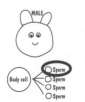

One sperm cell from the male unites with one egg cell from the female.

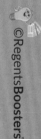

Fertilization
(Sexual Reproduction)

Remember that the body cell got split during meiosis?
Now, each gamete contains half the number of chromosomes found in a body cell

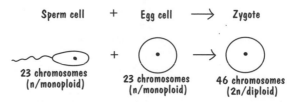

Sperm cell   +   Egg cell   ⟶   Zygote

23 chromosomes
(n/monoploid)

23 chromosomes
(n/monoploid)

46 chromosomes
(2n/diploid)

- **DIPLOID NUMBER:** Regular number of chromosomes of the species found in a **BODY CELL** (2n)

- **HAPLOID NUMBER:** <u>Half</u> the regular number of chromosomes of the species found in a **SPERM** or **EGG** cell (n)

- When the sperm (1n) and egg cell (1n) unite, they form a zygote (2n).
  1n + 1n = 2n

- A zygote is a **DIPLOID** (2n, which is the full number of chromosomes in a **BODY** cell)

Fertilization

(Sexual Reproduction)

The chromosomes of the gametes from the male and female combine.

The SPERM CELL (from the male) contains 23 chromosomes.
The EGG CELL (from the female) contains 23 chromosomes

23 + 23 = 46 chromosomes

The zygote will be formed with this combination. The zygote is only one cell, but it contains the full DNA instructions — how to create the body and make it run.

Body Cells vs. Gametes
(Sexual Reproduction)

- ✅ The diploid number in humans (body cell) = 46 chromosomes
  The haploid number in humans (gamete/sex cell) = 23 chromosomes

  Therefore, a human zygote contains 46 chromosomes

- ✅ The diploid number in frogs (body cell) = 26 chromosomes
  The haploid number in frogs (gamete/sex cell) = 13 chromosomes

  Therefore, a frog zygote contains 26 chromosomes

Internal Fertilization and Development

INTERNAL FERTILIZATION:

- The gametes meet inside the female's body.
- Increased chance the zygote will survive
  Therefore, less gametes are produced.

INTERNAL DEVELOPMENT:

- The development of the embryo happens inside the female's body.

People and mammals usually have internal fertilization and development.

External Fertilization and Development

EXTERNAL FERTILIZATION: When fertilization (the 2 gametes meet) happens <u>outside</u> the female's body

✅ Since birds have internal fertilization, they need to lay only a few eggs. However, fish, which have external fertilization, need to lay millions of eggs since the gametes are much less likely to survive. (They can freeze in cold waters, and they can be eaten by other organisms—Regents Question).

EXTERNAL DEVELOPMENT: The development of the embryo happens <u>outside</u> the female's body.

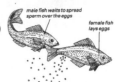

male fish waits to spread sperm over the eggs

female fish lays eggs

- If an organism had internal fertilization, there could be either internal or external development.
- If the organism had external fertilization, it must have external development.

Mitosis
(Sexual Reproduction)

Now we have a zygote, which is one cell that contains ALL the genetic information to make a baby.

But how does this one cell turn into a baby?

This one cell (zygote) copies itself over a lot of times with the process of MITOSIS (asexual reproduction).

Huh? Aren't we learning sexual reproduction (meiosis)? Yes! After meiosis happens to create the gametes, the gametes unite to form a zygote. After the zygote is formed, the cell undergoes mitosis (asexual reproduction) to turn into an embryo, which will eventually become a baby.

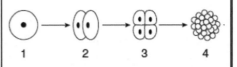

1
2
3
4

1. Meiosis created gametes which formed this zygote
2. Mitosis copied over the zygote
3. Mitosis copied over the 2 cells
4. Mitosis copied over the cells many times

Differentiation
(Sexual Reproduction)

Now we have a ball of a bunch of identical cells. How is this going to turn into a baby?

Another process happens now, called DIFFERENTIATION: The cells get specialized jobs now. Certain genes are expressed (turned on), while others are not expressed (turned off).

Differentiation: Each cell has a **different** job.

In skin cells, the genes for skin color are expressed, while the genes for eye color are not expressed.

In eye cells, the genes for eye color are expressed, while the genes for skin color are not expressed.

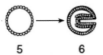

5. Differentiation: Cells start specializing in different jobs
6. This is a picture of a GASTRULA, which means the cells are becoming more specialized and developing into tissues and organs. This zygote becomes an EMBRYO during this stage.

As the embryo continues to develop, more differentiation takes place.

Differentiation
(Continued)

- Even though the full DNA is not expressed in every cell in the body, every cell in the body contains the complete set of DNA.

  > ✒ The cell in your eye has the "instructions" for creating your lungs, heart, liver etc.
  >
  > But, only the instructions dealing with the eye are actually carried out.

- Why does it happen? Certain genes are activated and others are deactivated, making different types of cells form.

- The environment has a big influence on how genes are expressed. (More on that later).

Growth
(Sexual Reproduction)

Now we have a gastrula with specialized tissues and organs. How does this become a baby?
Through the process of GROWTH

The Zygote will become an embryo, then a fetus and then a baby.

Baby ZEF: zygote, then embryo, then fetus

The fetus grows and develops in the UTERUS of a female. The fetus's needs are supplied through the PLACENTA.
(More on that later)

Fetal Growth From 4 to 40 Weeks

Meiosis Vs. Mitosis

MEIOSIS	MITOSIS
4 gamete cells produced from the original (even though in the female, only one egg is active)	2 cells produced from the original
Genetic variation from the original cell, the offspring has ½ of the genetic material of each parent	Final cells have same genetic makeup as the original cell
Exchange of genetic material between chromosomes (Recombination)	No exchange of genetic material between chromosomes
Purpose: to combine to create a zygote for reproduction	Purpose: growth/ replacement of body cells (in multicellular organisms)
2 cell divisions	1 cell division

Meiosis

2n

DNA Replication

2n

Meiosis I

1n   1n

Meiosis II

1n 1n 1n 1n

Mitosis

2n

2n

Mitosis

2n   2n

Problems in Meiosis

If the gamete contains more or less than half the diploid (23 chromosomes), there can be problems with the fetus.

✔ Human's diploid number is 46. Half the diploid number is 23. If a gamete contains more or less than 23 chromosomes, problems occur.

Effect:
• The zygote won't form -OR-
• It will produce an irregular individual

✔ Example: A Downs syndrome child will be produced if one of the gamete cells from the parent has 24 chromosomes instead of 23. (All together, the downs syndrome individual will have 47 chromosomes. 23+24 = 47.)

Female Reproductive System

1. Eggs are produced in the OVARIES.

2. OVULATION: The ovaries release an egg once a month which travels to the oviduct/fallopian tube. (Occurs on day #14 of the cycle.)

3. If sperm is present in the fallopian tube, it can fertilize the egg. The fertilized egg will then travel to the UTERUS where development of the fetus happens.

4. If there is no sperm present in the fallopian tube, the egg breaks down and hormones cause the lining of the uterus (blood and fluids) to be expelled. This causes MENSTRUATION.

5. The reproductive system is regulated by these HORMONES:

   • ESTROGEN: Regulates the menstrual cycle

   • PROGESTERONE: Maintains uterine lining during pregnancy

   • These hormones also affect puberty changes (hair growth, body development)

6. When women age, their menstrual cycle stops because they stop producing certain hormones. At this point, they can't have children anymore.

Female Reproductive System Diagrams

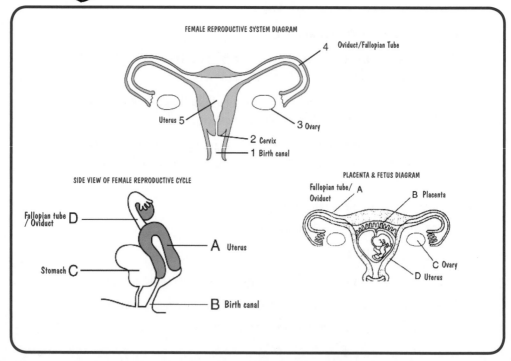

FEMALE REPRODUCTIVE SYSTEM DIAGRAM

4 Oviduct/Fallopian Tube

Uterus 5

3 Ovary

2 Cervix

1 Birth canal

SIDE VIEW OF FEMALE REPRODUCTIVE CYCLE

Fallopian tube / Oviduct D

A Uterus

Stomach C

B Birth canal

PLACENTA & FETUS DIAGRAM

Fallopian tube/ Oviduct A

B Placenta

C Ovary

D Uterus

Menstrual Cycle Diagram

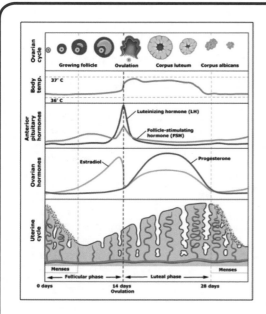

This diagram shows the different parts of the cycle (bottom line), and the hormones and body temperature at each part of the cycle. Recognize, but don't memorize.

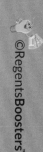

Placenta

PLACENTA: A structure that 1) brings nutrients and oxygen from the mother to the fetus and 2) helps remove its wastes (mostly carbon dioxide).

- Located on the side/ top of the uterus
- The umbilical cord attaches the fetus to the placenta.

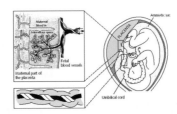

An expectant mother must be very careful with what she eats and drinks (and that she doesn't smoke) since these substances enter the fetus through the placenta.

❗ This is a common regents question.

Dangers to the Fetus

FETUS: The embryo becomes a fetus after 2 months and is called by this name till birth.

DANGERS TO THE FETUS:

Mother's dangerous activities	Effect: BIRTH DEFECTS in baby
Alcohol	Fetal Alcohol Syndrome (FAS)
Drugs	Brain damage
Smoking	Low birth weight

- These substances can be more harmful to the fetus than the mother because the fetus isn't developed yet.

- The most damage can occur from these dangerous activities in the first 3 months of the fetus' life because it's forming major organs then.

⬤ Toxins enter the fetus through the <u>placenta</u>.

Male Reproductive System

- Sperm (the male gamete) is produced in the TESTES (through MEIOSIS).

- Sperm is released from the testes through a tube (called the vas deferens).

- If the tube is cut/ injured, sperm will be produced, but not released.

- The male reproductive cycle is regulated by the TESTOSTERONE hormone which affects sperm production and puberty changes (voice changes, body hair)

Sperm Cell

1. Transports the genetic material (DNA) to the egg

2. ACROSOME: The head (top piece) of the sperm that contains the DNA, which has 23 chromosomes

3. NECK: Contains mitochondria which gives the TAIL energy (ATP) to travel to the egg cell.

   • A large amount of sperm cells are produced in order to increase the chance that one sperm cell will reach the egg.

● The egg cell doesn't move.

● Each sperm cell has different genes. Depending on which sperm cell enters the egg, the zygote will look different.

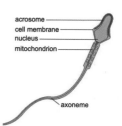

acrosome
cell membrane
nucleus
mitochondrion

axoneme

Sperm cell

Male Reproductive System Diagram

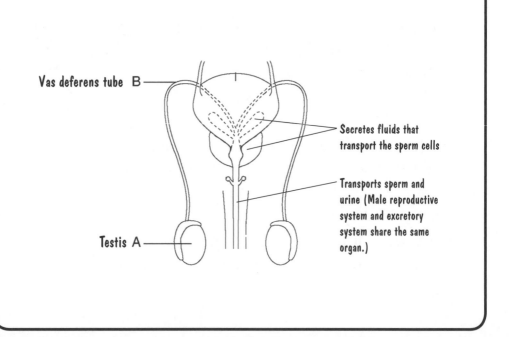

Vas deferens tube B

Secretes fluids that transport the sperm cells

Transports sperm and urine (Male reproductive system and excretory system share the same organ.)

Testis A

Characteristics of Sperm Vs. Egg

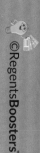

Characteristics of Sperm Vs. Egg

SPERM	EGG
Able to move	Not able to move
Smaller than egg	Larger than sperm
Many sperm cells produced	Only one healthy egg produced per month

Sexual Reproduction & Hormones

The human reproduction system is regulated by hormones.

✅ Examples of hormones used in the reproductive cycles:

Testosterone (males),

Estrogen & progesterone (females)

❗ This is a common regents question.

UNIT 4

GENETICS

DNA

DNA: The genetic information code

• Looks like a twisted ladder called a Double Helix ladder

• Made of repeating Nucleotides

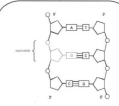

This DNA model contains 6 nucleotides (3 on right side, 3 on left side).

DNA Subunits & Pairing

- The DNA has 4 subunits (rungs on ladder):
    1. (A) Adenine
    2. (T) Thymine
    3. (G) Guanine
    4. (C) Cytosine

- (A) always pairs up with (T).
- (G) always pairs up with (C).

Therefore, a DNA molecule that has the linear sequence of <u>CTG ACA</u>, will have the opposite strand (complementary strand) of <u>GAC TGT</u>.

You don't need to memorize what each letter stands for.

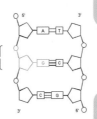

💡 <u>AT</u> - <u>A</u> pairs up with <u>T</u>.

Calculating the Percentage of a Specific Base

Sometimes, the regents asks you to figure out the percentage of either A, T, C, or G. Here's 3 rules you need to know and you'll be able to answer all such questions.

1. A+T+G+C=100% of the DNA base
2. A=T (If A=20%,T=20%)
   G = C (If G = 30%, C = 30%)
3. If you know the percentage of 1 letter, you can figure out the percentage of the rest of them.

> 📝 ex If A = 20%, Then T = 20%
>
> 20 + 20 = 40% (A+T)
> We need another 60% to add up to 100% and we have 2 bases left.
> C = G, so C= 30% and G = 30%
> C (20) + T (20) + C (30) + G (30) = 100%

DNA Making an Organism Look a Certain Way

DNA tells the body which protein or enzyme to make.

Every protein/ enzyme has a specific shape, which makes people's features different from one another.

✔ Everyone has different features (different shaped noses) because their DNA is different.

The reason children often have similar features to their parents is because they have similar (but not exactly the same) DNA as their parents.

DNA Structure

- The Sequence of the subunits (molecular bases) are <u>A,T,G,C</u>, which is the genetic code.

- The DNA sequence makes the body build different things (like skin, hair, blood, organs, enzymes etc.)

- If the order of the subunits changes, the item that is built will not be the same.

Genetic Information

- All genetic information (DNA, chromosomes, genes, and molecular bases) are stored in the **NUCLEUS** of a cell

- **CHROMOSOMES** are pieces of **DNA**

- **A GENE** is a small piece of a chromosome

- **MOLECULAR BASES**, which include A, T, C, G, are part of the gene

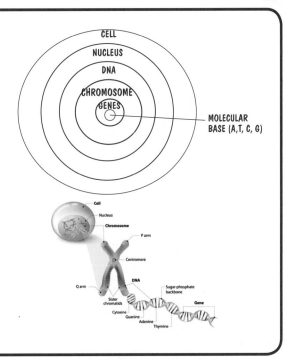

Levels of Organization for Genetic Information

You need to know which is the smallest unit, and which is the largest in genetic information terms:

<u>Largest to smallest:</u>
Cell
DNA
Chromosomes
Gene
Molecular Base (A,T,C,G)

<u>Smallest to largest:</u>
Molecular Base (A,T, C, G)
Gene
Chromosome
DNA
Cell

DNA Replication

How the DNA Replicates:

1. The double helix unwinds (molecule "unzips" the bonds) between the A&T or C&G.
2. Since there's a bunch of "single" nucleotides floating around, new nucleotides attach onto the old ones.

   ✅ New (C) attaches to Old (G).

⬤ The new DNA strands match the old ones. At the end, we have 2 matching double helix molecules.

Congratulations! Looks like a lot of singles are getting married!

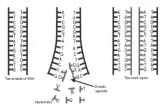

Two strands of DNA   Strands separate   Two exact copies

Nucleotides

Mutations/Problems in DNA Replication

MUTATION: A change occurred in one of the DNA bases. Something went wrong in the base of the sequence code.

> The order of the DNA got changed.

- Mutations in the DNA of <u>gametes</u> can be passed down to kids.
- Mutations in the DNA of <u>body cells</u> can NOT be passed down.

- Radiation and some chemicals can cause mutations like cancer.

- Usually, the mutation gives the organism a disadvantage.
- Some mutations can give the organism an advantage for natural selection.

Types of Mutations

1. **Deletion:** A molecular base gets deleted

   TAG becomes TA (The G got deleted)

   Or TAG becomes TG (The A got deleted)

2. **Insertion:** A molecular base gets inserted

   TAG becomes TCAG (The C got inserted)

   Or TAG becomes ATAG (The A got inserted)

3. **Substitution:** A molecular base gets substituted

   TAG becomes TCG (The C substituded the A)

   Or TAG becomes CAG (The C substituded the T)

Protein Synthesis

PROTEIN SYNTHESIS: Building a protein

How does a cell build protein?

<u>3 steps:</u>

1. DNA gets translated into the mRNA code (in the NUCLEUS)

DNA	⟹	mRNA
C	becomes	G
G	becomes	C
A	becomes	U
T	becomes	A

Looks like all the letters translate like DNA pairing. The only one who's different is the A becoming a U.

2. mRNA (messenger RNA) code travels to the RIBOSOME.
3. In the RIBOSOME, mRNA gets translated into an amino acid.
   You can figure out the amino acid by reading the Universal Genetic Code Chart (which will be given to you).

Universal Genetic Code Chart

Each mRNA code (3 letters) stands for another amino acid.

mRNA Codons and the Amino Acids They Code For

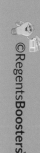

1. Make sure your code is in mRNA form, and not DNA form. If it's in DNA form, you must change it to mRNA form! (See chart on previous page.)
2. Find the mRNA code on the amino acid chart by finding the intersection between these three points:
   - The first letter of the mRNA code on the left side,
   - The 2nd letter on the top, and
   - The 3rd letter on the right side

   (If that makes you confused, just look for the 3 letter mRNA code randomly.)
3. The 3 letter code next to the mRNA code is the amino acid!

   GCU (mRNA) is (amino acid) Ala.   AAG (mRNA) is (amino acid) Lys.

Where Protein Synthesis Takes Place

- **NUCLEUS:**
  - Contains the DNA info (the instructions how to make the protein/genetic information)
  - Place where DNA gets copied into mRNA

- **RIBOSOME:** Place where the protein is made (protein synthesis)

Sounds like the nucleus has info to make a great product and outsources it to the ribosome factory.

The nucleus contains the important genetic info and the mRNA brings this message to the ribosome.

DNA vs. RNA

DNA	RNA
Double-stranded	Single-stranded
Never leaves the nucleus	Travels from nucleus to ribosome
Contains codes (codons) for making specific proteins	Contains anticodons (the opposite codes) of the DNA
Nucleotides are A - T & C - G	Nucleotides are A - U & C - G (U substitutes T, so if a DNA has an A, the mRNA will replace a U.)

How Amino Acid Affects Protein Function

1. DNA is translated to mRNA

2. mRNA is converted into amino acids

3. Amino acids determine the protein shape

4. Protein shape determines the protein function (shape specificity)

5. Proteins are synthesized for enzymes, eye color, nail growth etc.

Gene Expression

All DNA in the body are **IDENTICAL**. Each cell has a complete set of DNA with exactly the same instructions for making everything in the body.

But, only certain genes are expressed by certain cells, while the other ones are left quiet. That's how we have different types of tissue with different functions.

> In one place a hand will grow, in another spot, a nose will grow. Why? The hands genes are expressed where the hands grow, and all the other genes remain "silent" even though they are there. The nose genes are expressed where the nose grows, and all the other genes (including hands genes) are silent.

Gene Expression and the Environment

The environment can affect the expression of a gene.

✅ You were born with genes for light skin color. If you go out into the sun, your skin will change colors. Your genes didn't change. The way they were expressed changed because the environment changed.

✅ Plants have genes that tell them to produce chlorophyll. Therefore, they are green. If they don't get sunlight, chlorophyll cannot be made so plants will remain yellow/white.

🔦 This Is A Common Regents Question.

Two Types of Chromosomes

There are 2 types of chromosomes:

1. **AUTOSOMES:** Chromosomes that control everything in the body except the gender
    *   Autosomes make up 44 out of 46 of the human chromosomes. (22 out of 23 pairs)
2. **SEX CHROMOSOME:** Chromosomes that control the gender
    *   Sex chromosomes make up 2 out of 46 chromosomes (1 pair out of 23 pairs)

Autosomes are awesome. They control almost everything in our body!

Sex Chromosome

SEX CHROMOSOMES: Chromosomes that determine if the offspring will be male or female

Sex chromosomes make up the last <u>2</u> out of the 46 chromosomes (or the 23rd pair).

- Female's sex chromosomes: XX
- Male's sex chromosomes: XY

When the cells split during meiosis, each gamete has 22 autosomes and 1 sex chromosome.

- The female gamete will have 22 autosomes and one X sex chromosome.
- The male gamete will have 22 autosomes and one X <u>OR</u> Y sex chromosome.

Boy or Girl?

- The MALE chromosome influences the gender of the child.
  - ◆ The female sex cell will always be an <u>X</u>.
  - ◆ The male sex cell will either be an <u>X</u> or a <u>Y</u>.

- If a sperm cell containing an <u>X</u> chromosome unites with the egg, it will be a <u>girl</u>.

- If a sperm cell containing a <u>y</u> chromosome unites with the egg, it will be a <u>boy</u>.

💡 Boy ends with <u>y</u>. If the sex chromosome is a <u>y</u>, the child will be a boy.

Genetic Disorders

GENETIC DISORDER: A medical condition inherited by a child because of DNA abnormalities

Causes:

1. Mutation in the DNA sequence
2. Both parents are carriers for the disease

Carriers for Genetic Diseases

Someone who is a carrier will:

- **NOT** have the genetic disease themselves (or any symptoms)

- They can only pass the disease on to their offspring if <u>both</u> the father and the mother have the <u>same</u> genetic disorder.

> If both the father and the mother are carriers for Tay Sachs, their child can be born with Tay Sachs.
>
> If the father is a carrier for Sickle cell anemia and the mother is a carrier for Tay Sachs, their child has no danger of being born with either of these diseases.

Types of Genetic Disorders

1. **DOWN SYNDROME:** There is an extra chromosome in pair #21. The child can have delayed milestones and body differences (eyes, hand)

2. **SICKLE CELL ANEMIA:** Gene defect that causes red blood cells to be abnormally shaped. This causes blocked blood vessels, which causes the person pain, serious infections and organ damage.
   * Common in African American population

Normal Red Blood Cell   Sickled Red Blood Cell

3. **TAY SACHS:** Fatal, progressive destruction of the central nervous system. Slowly, the child becomes deaf and blind and the muscles stop working until he dies.
   * Common in the Jewish population originating from Eastern Europe

4. **HEMOPHILIA:** Blood doesn't clot normally

Genetic Counseling

Today, you can take a blood test to see which genetic diseases you are carrying. The male and female who wish to have children together both get tested for different genetic diseases.

- The counselors advise that you shouldn't marry someone who is a carrier of the same disease as you, so that your children won't have the disease.

- Hence, someone who is a carrier for Tay Sachs, can marry someone who's a carrier for Sickle Cell Anemia, but not someone else who's a carrier for Tay Sachs.

Genetic Tests

There are several tests to check for genetic diseases:

1. Blood test: A person can take a blood test to see if they're a carrier for a genetic disease or a child can get a blood test to see if they have a genetic disease

2. Amniocentesis: prenatal testing for chromosome abnormalities

    (This test is done in middle of pregnancy.)

3. Karyotyping: testing size, shape, and number of chromosomes

    (This test is done on the child.)

4. Electrophoresis (to be discussed soon...)

Selective Breeding

SELECTIVE BREEDING: Humans choose which plants and animals to breed.

- You can combine two types of plants and hope to get both of their good traits.

  ex Strong horses, domestic dogs, large eared corn

- But, watch out because you can also get the bad traits of both parents.

- You can increase **BIODIVERSITY** (variety of life on earth) by creating organisms with new characteristics.

- (But, you can decrease **BIODIVERSITY** with selective breeding too! If you keep breeding a specific type of dog, the other types of dogs will decrease.)

Genetic Engineering

GENETIC ENGINEERING: Using technology to change organisms

Also known as:
- GENE MANIPULATION
- BIOGENETICS

- Genetic engineering is used for:
    1. Creating insulin and other hormones
    2. GENE THERAPY: Curing illnesses by replacing defective DNA in a sick person's body with healthy DNA
    3. Creating better crops (pest resistant, sweeter, larger)

Gene Splicing

GENE SPLICING: Cutting and replacing DNA coding

1. Cut off part of the DNA code
2. Replace it with a different code

> If there's a part of a chromosome you don't like, you can cut it out.

Sounds like cutting and pasting.

Scientists are working on cutting out the faulty chromosomes in a person who has a genetic illness (like Sickle Cell Disease), and replacing it with healthy chromosomes, thereby curing them of the illness!

This is a form of genetic engineering.

Example of Genetic Engineering: Insulin

Insulin and Genetic Engineering

<u>Until now:</u>
Diabetics who didn't produce enough insulin had to get their insulin from sheep and horses.
• CONS: Animal insulin can cause allergic reactions, and is expensive.

<u>Using genetic engineering today:</u>
Get insulin from genetically modified bacteria, which can produce HUMAN (also called SYNTHETIC) insulin.
• PROS: Cheaper, less allergic reactions

How To Produce Human Insulin

bacteria ring

insulin DNA
Bacteria ring

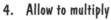

Human Insulin

1. Cut off the non-essential part of a nonharmful bacteria ring

2. Cut off the "insulin instructions" from the DNA of a healthy body cell (from a donor)

3. Fuse the bacteria ring (with empty space) + "insulin instructions" from DNA together
   - We place the "insulin instructions" DNA into the bacteria ring because bacteria reproduces very quickly and we want to make the "insulin instructions" DNA reproduce quickly.

4. Allow to multiply

5. Diabetic now has insulin to lower his or her blood sugar level!

Notes on Producing Human Insulin

- This process is complicated and it is very hard to get it to work. But once you get it to work once, you hit the jackpot! The cell can now reproduce using mitosis (making the exact same copies) so that all future bacteria that result from the altered one will also produce human insulin.

- This produces human insulin that a diabetic can use to inject into himself with, but doesn't cure him or make him start producing his own insulin.

  The only way to cure a diabetic is through a pancreas transplant or replacing the defective gene.

Gel Electrophoresis

GEL ELECTROPHORESIS: Process of separating molecules based on size and charge

- Separates DNA and RNA molecules by size using electric current. The electric current causes the DNA fragments to move.

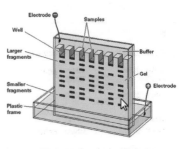

- How does it get separated? Smaller molecules move faster and move farther down than larger ones.
- The machine has a negative and positive end. Molecules of the DNA are charged and move to the opposite pole.

Uses of Gel Electrophoresis

- Way to prove or disprove if someone did a crime

- Way to prove or disprove paternity (if someone really is a child's parent)

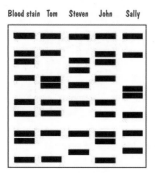

Blood stain   Tom   Steven   John   Sally

These are gel electrophoresis results. We see the pattern of DNA of the blood stain on the left column. Then we see the pattern of DNA of Tom, Steven, John, and Sally. Which person's DNA matches the blood stain? He's probably the culprit!

Problems with Genetic Engineering

- It's difficult to figure out the location of genes for specific proteins.

- Inserting genes into humans from other species raises ethical questions: Should humans be manipulating genes?

- It may produce toxins that can harm us (especially with GMOs: genetically modified organisms) when used in foods.

Pros of Genetic Engineering

- **GENE THERAPY:** Transplanting normal genes into cells in place of missing or mutated genes

- There would be more food for the world if crops are genetically engineered and able to resist pests

- We could produce hormones that humans are missing (like insulin)

- We can produce medicines and vaccines

I wish I could tell my friends to stop their psychotherapy and go for gene therapy instead, and get happy genes (like mine)!

Resistance

RESISTANCE: The ability to remain unchanged by the damaging effect of something

- All organisms have different genes even within the same species. That is why you and your friend look different even though you're both human. Some organisms will be born immune to elements.

  🔅 Someone may be born immune to the flu. (No matter how many times they are exposed to it, they will not get sick.)

- Genetic advantages help organisms survive in life and death situations. If something causes most of a species to die out, an organism that has a genetic advantage or resistance may be able to survive. If it survives long enough to reproduce, its offspring will probably be resistant too.

Example of Resistance with Bugs

Two bugs out of a group of 100 bugs were born with a genetic variation that made them resistant to pesticides. When the pesticide is sprayed, these 2 bugs are the only survivors, and the rest of the bugs die. These 2 bugs will probably live long enough to reproduce and most of their kids will be resistant to the pesticide.

- When farmers use pesticides, they are making themselves problems for the future because most of the offspring of the 2 surviving bugs will also be resistant to the pesticide. So, the farmer will use even stronger pesticides and the process will repeat itself until SUPER BUGS are created: bugs resistant to pesticides.

Example of Resistance with Antibiotics

- When a person takes antibiotics, it kills out the weak bacteria, but can leave some resistant bacteria (bacteria that didn't get killed out from the antibiotics).
- These resistant bacteria will reproduce and you'll need stronger antibiotics to kill these tougher bacteria out.
- Later on, this process can repeat itself until you are left with such tough bacteria that antibiotics can't destroy them.

So, only take those antibiotics under your doctor's instructions.

UNIT 5

ECOLOGY

Note: There are many, many regents questions on this topic. Make sure you know this topic extra well. Specifically, there will be many long answer questions on the topic of human impact on the environment. Use your brains and common sense! If you give a correct answer that you didn't learn, it will be marked correctly.

Ecology

ECOLOGY: The study of how organisms interact and depend on other organisms and on the environment

Biotic and Abiotic Factors

- **BIOTIC FACTORS:** <u>Living</u> organisms

  ✅ Fish, snails, plants

- **ABIOTIC FACTORS:** Non-living factors (that are important to living things)

  ✅ Water, temperature, sunlight

Abiotic. <u>A</u> means <u>not</u>. Abiotic means <u>not</u> living.

Habitat

HABITAT: The organism's "address"/place where they live

✅ The habitat for a bear is a cave.

✅ The habitat for birds is a nest in a tree.

An organism will go to the habitat that best suits his needs — the right temperature, amount of water etc.

Limiting Factors

LIMITING FACTOR: An abiotic or biotic factor that limits the survival and growth of an organism

- Sunlight, plants, water, temperature etc.

- If there wasn't enough sunlight, plants would die. In this case, the sunlight would be the limiting factor.

- In an aquatic (water) ecosystem, the saltiness of water would be the limiting factor.

• Some resources are finite/ limited, so organisms cannot reproduce in an unlimited way.

Limiting Factors Graph

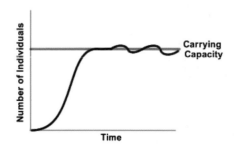

A population will grow until it reaches its **CARRYING CAPACITY**: the maximum the ecosystem can hold. The carrying capacity is the line across the graph.

The population is limited by limiting factors (sun, water, food).

OVERPOPULATION: The population grows above the carrying capacity

UNDERPOPULATION: The ecosystem can hold more of the population

Biome

BIOME: A very large area, with a specific climate, and specific types of plants and animals

> Types of biomes: desert, grassland, tropical rain forest

Ecosystem

ECOSYSTEM: Specific place where all living and non-living factors live and interact with each other

Pond, forest, aquarium

- Humans, animals, and plants all live in an ecosystem.

Ecosystem for fish: water, other fish, seaweed, air, rain

Population / Community / Ecosystem/Biome

What is the difference between a population, community, ecosystem, and biome?

POPULATION: All organisms of one species that live in the same place at the same time

> All the rosebushes in one garden (but not all the other kinds of flowers or animals)

COMMUNITY: All biotic (living species) in an area

> All the rosebushes, daisies, grass, bushes, mice, bugs etc. in one place, but not the abiotic factors

ECOSYSTEM: all biotic + all abiotic factors in a given area

> All the rosebushes, daisies, grass, bushes, mice, bugs in one place & air, soil, temperature, light

BIOSPHERE: Place on earth where life is found

Levels of Organization in an Ecosystem

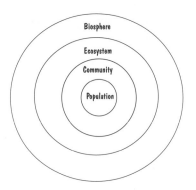

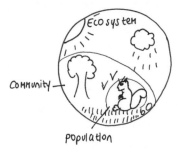

The population is part of the community.
The community is part of the ecosystem.
The ecosystem is part of the biosphere.

Symbiosis

SYMBIOSIS: A close relationship between organisms

🔍 Symbio**sis**: **sis**ters
have a close relationship
(hopefully!).

Types of Symbiosis

1. **MUTUALISM: Both organisms gain**

   ✅ Humans give carbon dioxide to plants, while plants give oxygen to humans.

2. **COMMENSALISM: One organism gains, while the other doesn't gain or lose.**

   ✅ Crab inside oyster's shell — the crab gets shelter from the shell but doesn't harm it

3. **PARASITISM: One organism gains (PARASITE), the other loses (HOST).**

   ✅ Pinworm inside human body

Parasites/Host

- **PARASITE:** An organism that lives off another organism

  ◆ The parasite gets its food and shelter from the organism.

  ◆ The parasite harms the organism, but doesn't kill it.

  ⌖ Tapeworm in human intestines: He gains a place to live and food to eat, but the human is harmed.

  🔘 <u>Parasite</u>. <u>Party</u>. At a <u>Party</u>, you eat, drink, and stay in someone else's house / hall. A parasite has a party off whomever he's living in!

- **HOST:** Hosts the parasite living inside its body

  ◆ This guy is usually harmed but not usually killed

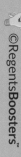

Producers and Consumers

- **PRODUCERS:** The organims that produce their own food

  ✔ Plants and algae

  - They produce food (make their own food) through photosynthesis
  - Producers have the most energy because they get their energy directly from the sun

- **CONSUMERS:** Organisms who eat producers or other consumers
  - Consumers cannot produce their own food! Why? People and animals cannot do photosynthesis.
  - Types of consumers: Herbivores, carnivores, decomposers

  ✔ Mice eat corn, snakes eat mice

💡 <u>Producers</u> <u>produce</u> food.

💡 <u>Consumers</u> <u>consume</u> food.

Herbivores

HERBIVORE: Animals that only eats plants

PRIMARY CONSUMERS: These are the first consumers on the chain to get food or energy from the producers.

🗝 Herbivores.
<u>Herb</u>s are plants.
<u>Herbi</u>vores eat plants.

✅ Chickens, cows

Carnivores

CARNIVORE: Animal who eats other animals

- They are considered **SECONDARY CONSUMERS:** They don't get their energy directly from the producers. They eat other consumers.

 Lion eats meat.

💡 <u>Car</u>nivore. <u>Crie</u> = <u>Cry</u>. When an animal eats another, the animal being eaten <u>cries</u>.

Scavenger/ Prey & Predator

There are two types of carnivores (animals that eat other animals):

1. **SCAVENGER:** Organism that gathers remains of a <u>killed</u> animal

> ❶ <u>Prey</u>. <u>Pray</u>. <u>Prey pray</u> that they don't get eaten.

2. **PREDATOR:** Organism that hunts, kills and eats other <u>live</u> animals

   PREY: The animal that gets hunted

Prey & Predator Graph

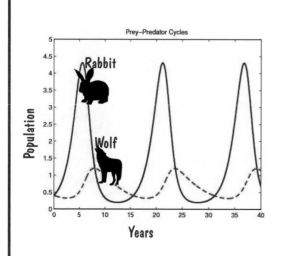

Prey–Predator Cycles

- The rabbit is the prey and the wolf is the predator.

- The rabbit population is able to reach a higher population than the wolf population. Why? Because rabbits eat producers, which have the most energy.

- When the wolf population grows, the rabbit population gets smaller because the wolves are eating them up!

- When the wolf population is low, the rabbit population grows because the wolves aren't eating them.

Decomposers

 ✔ex Fungus, bacteria, worms, mushrooms

DECOMPOSERS: Organisms that break down dead or rotten organisms

✔ex Bacteria breaks down a dead horse

- Decomposers return **NUTRIENTS** (not energy) to the environment.
- Decomposers are the last organism on the feeding chain.

Cemeteries have extremely fertile earth since the nutrients from the dead bodies get absorbed into the soil with the help of decomposers.

Trophic Levels

TROPHIC LEVELS: Eating levels

Level 1: Producer

Level 2: Consumer (herbivore)

Level 3: Consumer (carnivore)

Level 4: Decomposer

Producer: Grass →

Herbivore: Grasshopper →

Carnivore: Bird →

Carnivore (#2): Bobcat →

Decomposer: Bacteria

🕐 Trophic Levels.
Trophy: Successful people get trophies. The organism that finds what to eat is successful.

Food Chain

FOOD CHAIN: The direct transfer of energy from one organism to the next

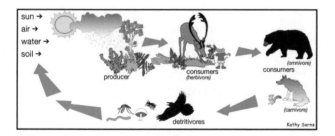

sun →
air →
water →
soil →

producer

consumers
(herbivore)

(omnivore)
consumers

(carnivore)

detritivores

Kathy Sarns

The arrows point to the organism whose stomach the food is going into. The deer is going into the bear's tummy. The tree is going into the deer's tummy.

⬤ Notice the directions the arrows are pointing to. The arrow is pointing toward the organism that eats it. These are common regent questions.

• The food chain gets its energy from the sun.

Removing Organisms in a Food Chain

Food Chain
Arrow points to organism that ate it

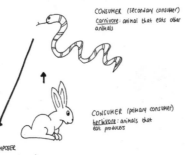

CONSUMER (secondary consumer)
<u>Carnivore</u>: animal that eats other animals

CONSUMER (primary consumer)
<u>herbivore</u>: animals that eat producers

DECOMPOSER

PRODUCER (autotroph)

The rabbit eats the grass.
The snake eats the rabbit.
The decomposer (worm, mushrooms) eat the snake.

If all the rabbits died out in this ecosystem, the snakes would die out because they wouldn't have food.

The food chain can be shown in different ways:

1. In a picture (see image)

2. Horizontal:
   Grass → Rabbit → Snake → Worm

3. Vertical:
   Grass
   ↓
   Rabbit
   ↓
   Snake
   ↓
   Worm

Food Web

FOOD WEB: The complex, interconnecting food chains in a community

- ● Notice the direction of the arrows. Once again, the arrows are pointing toward the organism that eats it.

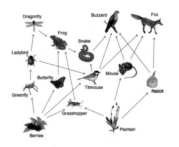

- • The food web gets its energy from the sun
- • The food web is much more realistic than the food chain since each organism usually eats more than one food.
- • If one organism is removed from the food-chain, other organisms get affected.

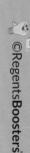

Food Chain Vs. Food Web

FOOD CHAIN is much more simplified and less accurate than the **FOOD WEB**.

🔵 Common Regents Question: What will happen if one organism disappears from the chain/web?

- If it disappears from the <u>chain</u>, the organism that eats this organism will die (which isn't the most accurate in real life).

- If it disappears from the <u>web</u>, the organism who eats it will find an alternate choice, but there will be a decrease in that organism, and there will be an increase in the food of the organism that disappeared from the chain.

Removing or Adding an Organism to the Food Web

Examples of how removing or adding an organism to the food web can affect other organisms:

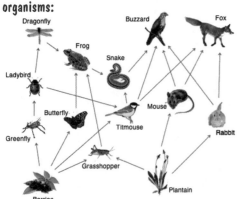

- If you introduce beetles (who eat berries) to the food web, the butterfly population will decrease since it will have less food.

- If the plantain is removed, the rabbit will die out, which will cause the fox population to decrease because he usually eats the rabbit.

- If you remove the berry plant, the grasshopper population will decrease, but not die out because he can still eat the plantain.

- If you remove the mouse population, the fox population will decrease and the rabbit population will decrease because the fox will eat more of them now.

Energy Pyramid

ENERGY PYRAMID: Shows how the sun's energy passes through the food chain

FINAL CONSUMER: Gets the least energy

CONSUMERS (CARNIVORE): Gets less energy than herbivore consumer

CONSUMERS (HERBIVORES): Gets less energy than producers

PRODUCERS: Gets the most energy from the sun

- Producers receive the most energy from the SUN through the process of photosynthesis and have the most available energy to pass on.
- As the food moves up the food chain, energy is lost.
- Energy is recycled in the food chain and food web.

Pyramid of Biomass

There are the most producers, less consumers (herbivores), even less consumers (carnivores), and the least final consumers. The higher up the pyramid, the fewer organisms there will be.

Read this card together with the next card's diagram.

- **WHY?** The producers have the most energy from the sun, therefore there are the most producers.
- The final consumers have the least energy, therefore there are the least final consumers.

> ex There are many more blades of grass than grasshoppers, and many more grasshoppers than lions.

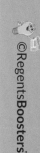

Pyramid of Biomass Diagram

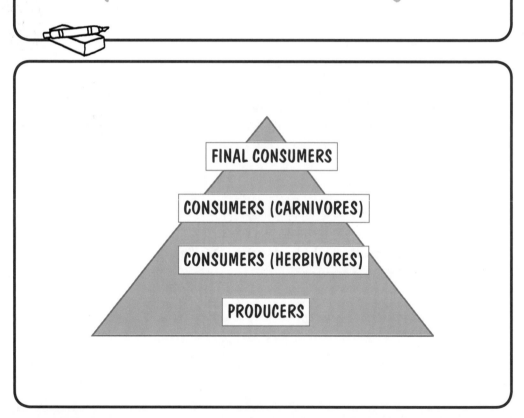

Pyramid of Energy & Biomass

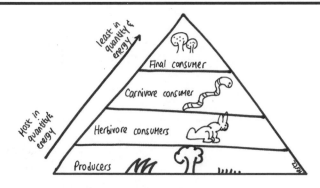

- Food chains and food webs get energy from the sun.
- The lower an organism is on the pyramid, the more energy they have. Producers have the most energy because they get energy directly from the sun.
- As you move higher on the pyramid, there are less of each population because they have less energy because they are farther away from producers who have the most energy.
- As you move higher on the pyramid, there is less energy because energy is lost <u>as heat</u> into the environment.

🛈 This is a common regents question.

Recycling in the Ecosystem

There are three main recycling cycles in the ecosystem:

1. <u>C</u>arbon <u>O</u>xygen Cycle

2. <u>W</u>ater Cycle

3. <u>N</u>itrogen Cycle

People have the ability to disrupt these cycles by destructive behaviors (like pollution).

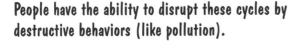
💡 The first letter of each cycle spells <u>COWN</u>.

Carbon-Oxygen Cycle

Carbon-Oxygen Cycle

1. Carbon is taken from carbon dioxide in the air.

2. Plants use carbon dioxide for photosynthesis and to make their glucose.

3. Animals use oxygen for respiration from the air, which comes from plants that do photosynthesis.

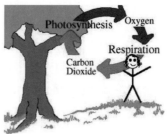

4. Animals breathe out carbon dioxide, which gives the plants ingredients to do photosynthesis.

Problems in the Carbon-Oxygen Cycle

COMBUSTION: The process of burning something

Carbon dioxide is released when fires are burned, or machines are used. Oxygen is also used up.

This can lead to problems like global warming.

- Since forests are cut down in order to build factories, there are no trees available in those areas to consume the carbon dioxide and produce oxygen.

Biodiversity

BIODIVERSITY: Variety of living things

- Extinction lowers the biodiversity levels. (If there is less of an organism, there is less variety.)
- Why we want biodiversity:
  When we have a large variety of all different types of organisms, we can get food, medicines, insecticides, and many other useful resources from them.
- If you pollute a river, you are causing the biodiversity to decrease since it will cause the extinction of some types of fish (which means you'll have less food and medicines).
- ❗ There are many regents questions on this topic.

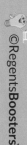

Niche

NICHE: The specific role an organism plays when it interacts with its environment

- The more niches available, the more genetic variation there will be.

> Variations in beaks of Galapagos Island Finches show that there were many niches available on those islands and many different birds with different types of beaks were able to survive by eating different types of seeds.

- If dramatic changes happened in the island's environment, and there's competition with several organisms over the same niche, natural selection can occur and only the fittest (those who adapt best) will survive and reproduce.

Stable Community

STABLE COMMUNITY: A community where the organisms are able to live for a long time with very little change

- A stable community contains:
  1. Biodiversity: many different types of organisms
     This way, if one organism dies out, it won't affect the other organisms too much because there's so much other food available.
     However, if there's little biodiversity, if one organism dies out, it can seriously harm the entire ecosystem and make it difficult for it to recover.
  2. There are more producers than consumers.
  3. The community contains producers, consumers, and final consumers.
- If the community is able to last for so long, there will be little evolutionary change.

Ecological Succession

ECOLOGICAL SUCCESSION: An ecosystem changes very gradually

- The ecosystem starts out as some small plants, then gets shrubs, then small trees, then big trees until it becomes a stable community
- When the shrubs start growing, the grasses decrease, when the trees start growing the shrubs decrease...
- Usually takes hundreds or thousands of years

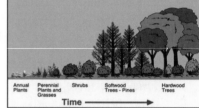

Annual Plants | Perennial Plants and Grasses | Shrubs | Softwood Trees - Pines | Hardwood Trees

Time ⟶

> ✐ It takes many years for a field of grass to change into a forest when starting from bedrock

- When natural disaster strikes, the succession will happen real quickly and will eventually look identical to the original ecosystem.

> ✐ If an earthquake or tornado occurs in a forest, the ecosystem can recover and rebuild itself after the disaster faster than if it was being built from scratch.

❗ There are many regents questions on this topic.

People Ruining the Environment: Deforestation

DEFORESTATION: Cutting down forests

Problems:

1. More carbon dioxide in the air because there's less trees to do photosynthesis and produce fresh, clean oxygen

2. Makes organisms who live off these trees lose their habitat and die

3. Less biodiversity

Solutions:

1. Replant the trees that you cut down

2. Recycle more that we don't need to cut down so many trees

People Ruining the Environment: Direct Harvesting

DIRECT HARVESTING: Destroying specific species

 Killing the mosquitos in your garden

Problems:

1. Lowers the biodiversity, which means less food and medicine for us

2. Makes ecosystems less stable — Removing one organism can affect the other organisms

Solutions:

1. Limit hunting and farming

2. If you don't want a pest in a certain area, relocate the pest to another area

People Ruining the Environment: Importing Species

Importing plants or animals from other countries causes problems:

1. The plant or animal can carry diseases with them

2. The new plant or animal can disrupt the ecosystem, taking away food from an existing organism

Solutions:

1. Don't import plants or animals

2. Pass laws to inspect imported plants and animals

People Ruining the Environment: Overfarming

OVERFARMING: Planting the same crop on the same piece of land

- The topsoil (the top level of the soil) is filled with many nutrients.
- When farmers use the same plot of land to plant the same crop over and over, they are depleting the earth of nutrients.

  - There is a tradeoff (an exchange of one thing in return for another) here. The more the farmers farm, the more food they produce and the more money they make. On the other hand, they are ruining the environment.

  If farmers plant tomatoes over and over and over on the same plot of soil, the soil will no longer have the nutrients it takes to produce tomatoes.

<u>Solution:</u> Farmers should rotate planting different crops on different parts of the soil.

People Ruining the Environment: Overgrazing

OVERGRAZING: Animals graze on the same soil over a long period of time

- Allowing animals to graze on the same part of soil over a long period of time will remove the grass holding down the topsoil (richest layer of soil). Now the soil can **ERODE**: wash away in the rain/ blow away in wind.

<u>Solution:</u>
Rotate the plots of land the animals graze on and don't let them completely eat up the grass in one area.

People Ruining the Environment: Pesticides

Pesticides are chemicals that kill out pests.

Problems:

1. Killing pests lowers biodiversity
2. Pesticides contain dangerous chemicals that can cause cancer and diseases

Solutions:

1. Use traps to kill insects
2. Use pest-eating plants
3. Use pest resistant plants (plants that pests don't eat)

People Ruining the Environment: Natural Resources

NONRENEWABLE RESOURCES: Resources that cannot be replenished if they get used up

> Oil/gas/coal/fossil fuels, stones, gold, silver, diamonds, minerals

Problem: People are using these resources constantly. Once they are used up, we cannot replenish them!

The gas we use to fill up our car is sometimes called: oil, coal, or fossil fuels

Solutions:
- Don't use the nonrenewable resources too much
- Find alternative energy sources
- Use solar powered cars (cars that don't need gas)
- Use public transportation, instead of driving your car that uses gas

People Ruining the Environment: Polluting the Water

Types of water pollution:

1. Regular water pollution: Many factories dump their toxic chemical wastes into the water. → These poisons kill the fish and can be harmful to people and animals who drink the water.

   • Gardeners often use fertilizers to help plants grow. When it rains, the water drags some fertilizer into the water. This increases the water's PH level, which can cause fish to die.

2. THERMAL POLLUTION: Industries dump heated water or air into the water, which lowers the oxygen levels, and kills fish

• Both regular water pollution and thermal pollution decreases the oxygen level in the water, which helps the fish breathe. If the fish can't breathe, they die. → This decreases the biodiversity of the water population →less food and medicine

Solutions for Water Pollution

- Factories should work on producing machinery that produces less toxic waste.

- Gardeners should use less fertilizer.

- We should reuse manufactured products that the factories won't have to use their machinery (thus polluting the water) to create more of the product.

People Ruining the Environment: Polluting the Air

- **POLLUTANTS:** A substance that pollutes, making the environment less fit for living things
- **COMES FROM:** Burning **FOSSIL FUELS** (coal, oil, natural gasses)

  🔧 Cars, heat, machinery in industries.

- **CAUSES:**
  - ◆ Acid rain
  - ◆ Global warming

- <u>Solution:</u> Find another non-polluting source of energy.

  🔧 Solar Energy: Getting energy from the sun

Global Warming

- When there is too much carbon dioxide in the air (caused by burning fossil fuels), the gas forms a layer around earth that traps in heat and doesn't let the heat escape.

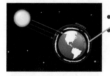

The earth has a layer of pollution around it. The sun's heat enters the atmosphere, but cannot escape because it's trapped in by the pollution gasses.

- This can cause the earth to get too warm.
- We are concerned that the earth may become very hot, melt glaciers (massive blocks of ice), and flood the earth.
- Air pollution is sometimes called GREENHOUSE GASSES. This is because the gasses trap in heat, just like a greenhouse traps in heat.

🔵 There are many regents questions on this topic.

Acid Rain

- When fossil fuels are burnt, chemicals are formed which rise and attach themselves to the clouds.
- When it rains, the water gets mixed with these chemicals.
- The rainwater (which is supposed to be neutral on the PH scale) becomes acidic.
- These rains can destroy forests, crops, and fish (lowering biodiversity), and are unhealthy for us to drink.

Gasses rise to the surface of the clouds. When it rains, the gasses combine with the water droplets to make acid rain.

People Ruining the Environment: Ozone Depletion

- There is a layer of gas that surrounds the earth, called the ozone layer.

- The ozone layer stops harmful sun rays (UVA and UVB radiation) from entering our atmosphere. (The ozone layer is a "good guy".)

- Certain pollutants (like chemicals found in aerosol spray deodorants and air fresheners, and exhaust from air conditioners, etc.) make this layer go away (get depleted).

- Once the ozone layer is destroyed, UV Radiation, harmful sun rays, can enter the atmosphere and cause skin cancer.

Ozone layer

Solutions for ozone depletion:

- Find substitutes for spray cans that deplete the ozone layer. (Like stick deodorant, and non spray air fresheners)

- Create machinery that produces fewer harmful gasses.

- Reduce the use of substances that pollute the air

People Ruining the Environment With Their Garbage

1. Humans waste a tremendous amount of natural resources. (Once it's gone, it can't come back!)

2. Wastes often contain harmful chemicals, which can pose a health threat when stored in landfills.

3. We are running out of space in our dumps.

Overpopulation

OVERPOPULATION: Population of a species exceeds the carrying capacity of its ecological niche

- Most species reproduce a little more offspring than the environment can hold (exceeds carrying capacity by a small amount).

- Within a short time, the population returns to numbers within the carrying capacity because some organisms die because of limited resources (not enough food, water, or air).

📝 A certain tree can hold (has a carrying capacity of) 500 birds, and after a while 500 birds actually live in this tree. Then, another 100 are born, totaling 600 birds. 100 birds (not necessarily the newborns) will die or have to move away because there isn't enough food and space for them.

Overpopulation and the Environment

- Scientists worry that since the population of humans is growing so rapidly, the limited resources will run out, and there won't be enough natural resources for people.

- Scientists say that overpopulation causes all the other problems in the environment: deforestation, ozone depletion, global warming... They feel that humans created more destruction than anything else. Therefore, some advocate population control. Others disagree with this approach since the world was ultimately created for people.

Alternative Energy Sources

Many problems in the environment come from using machinery and gas. We need to try to find alternative energy sources to minimize the damage we are making on our environment.

Alternative energy sources:

1. **SOLAR POWER: Energy from the sun**
   - Pros: no pollution, renewable
   - Cons: If there's no sun, there's no energy, can be expensive

2. **WIND POWER: Energy from the wind**
   - Pros: no pollution, renewable
   - Cons: If there's no wind, there's no power

3. **COMPOST ENERGY: Energy from compost (breaking down organic waste and using it   as an energy source)**
   - Pros: No pollution, there's plenty of organic garbage so we won't run out of it
   - Cons: May cause rodent problems, it's difficult for people to dispose organic materials separately and it stinks!

NYC Compost
Garbage Bin

🔵 There are many regents questions on this topic.

Alternative Energy Sources

4. **WATER ENERGY (HYDROELECTRIC):** Getting energy from the water

- **Pros:** renewable, reliable, less air pollution

- **Cons:** can harm fish, expensive, if there's a drought — there won't be water energy

5. **NUCLEAR ENERGY:** Energy created from nuclear reactions

- **Pros:** Less pollution, reliable

- **Cons:** Dangerous chemical waste, nuclear accidents cause major catastrophes, nonrenewable

Solutions to Saving the Environment

Solutions to Saving the Environment:

1. Reduce

2. Reuse

◉ The 3 R's:
Reduce, reuse, and recycle.

3. Recycle

4. Pass laws to stop environmental damage

5. Create RESERVE PARKS: parks that protect important wildlife

6. Use alternative energy sources (solar, wind, compost, nuclear or water energy)

Tradeoffs

TRADEOFF: A situation where you must choose between two opposite things

- Humans use natural resources, cut down forests, use gas cars, use pesticides etc. for good reasons
- However, there are major disadvantages to these actions
- Therefore, people need to weigh the pros and cons of each action to decide what to do

ACTION	PROS	CONS
Using gas	Creates jobs, makes our cars run	Pollution
Deforestation	Use wood for furniture and paper	Decreases biodiversity
Pesticides	Stop pests from eating our plants	Harmful chemicals

In each situation we need to gather information on all aspects and evaluate the situation

UNIT 6

EVOLUTION

Evolution

- <u>**DARWIN**</u>: Father of evolution

- Darwin states: Life started billions of years ago with simple, unicellular organisms. Species evolved (changed) over time to become more complex through **NATURAL SELECTION**

- EVOLUTION: Change in the characteristics of a population over time

Natural Selection/ Survival of the Fittest

NATURAL SELECTION/SURVIVAL OF THE FITTEST: The best adapted organisms survive and reproduce

- All species fight for limited natural resources (COMPETITION).
- Organisms that ADAPTED (changed to become more fit) win the competition.
- These ADAPTIVE ORGANISMS (organisms that adapted) will survive and reproduce.
- Organisms who cannot adapt, die and become extinct.
- Therefore, evolutionists say that the population improves over time because there are more adaptations (EVOLUTIONARY CHANGES).

❶ In a game of ball, everyone is <u>competing</u> over the ball (natural resources). Whoever has the best <u>adaptive trait</u> (knows how to catch, throw, dodge, and duck) will get the ball and <u>stay in</u>! (They have the natural traits to survive.)

Adapted Organisms

- The organisms that adapted best to the environment will most likely survive and pass on their good characteristics to their offspring. Hence, these characteristics will become more and more common with each generation.

- On the other hand, organisms that have characteristics that lessen the chance the organism will survive and reproduce, will lessen in amount over time.

- Eventually, almost all the organisms in a given population will have that good characteristic.

Extinction

EXTINCTION: Species that cannot adapt to the environment die out and become extinct.

- More than 99% of all species became extinct over time

- **ENDANGERED SPECIES:** Species at risk of becoming extinct

- We need to try to minimize extinction because we want more biodiversity. This can be done by making laws against hunting and against destroying natural habitats.

Extinct Mammoth

Adaptations

ADAPTATIONS: Characteristic of an organism that helps it survive in difficult circumstances

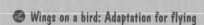

Wings on a bird: Adaptation for flying

All types of adaptations in evolution occur in the DNA and are not caused by learned behaviors.

Types of Adaptations

1. STRUCTURAL CHANGES: Changes in the structure of the organism

> Wings on a bird, color of a bug (camouflage)

2. BEHAVIORAL CHANGES: Changes in behavior
Males and females attract each other through different actions (courtship displays)

> Animals dance and make sounds to attract a mate.

3. FUNCTIONAL CHANGES: Changes in the way the organism functions

> Poison of a snake

All these types of adaptations give an organism an advantage and help it survive.

Stable Vs. Unstable Ecosystem

- If the environment changes, more evolution will occur because organisms will need to adapt in order to survive.

- If the environment stays the same (stable ecosystem), very little evolution will occur.

How do Adaptations Occur?

1. **SEXUAL REPRODUCTION:** During sorting and recombination of the genes, new traits are created

2. **MUTATIONS:** A change happens in the DNA
   - The mutation can be a disadvantage, which will make the organism die out.
   - The mutation can be an advantage, which will help the organism survive.
   - If the mutation is in a body cell, children cannot inherit it. If the mutation is in the gamete, children can inherit it.

Mutation: mute. A <u>mute</u> person is different than a regular person. Similarly, a <u>mutation</u> will make the gene different.

Rate of Evolution

Organisms with short reproductive cycles and many offspring (such as bacteria) evolve quicker than those with long reproductive cycles and fewer offspring.

Why? So much reproduction is happening, which increases the probability that a mutation will happen.

Bacteria

Genetic Variation

GENETIC VARIATION: Differences in genes amongst offspring

> Within the butterfly species, some are dark yellow, others are light blue. Even though they have different genes, they are still part of the same species.

- The more different species with variations there are, the greater the biodiversity.
- The greater biodiversity makes it more likely that someone will survive a disaster (such as a flood or volcano) because it will have the adaptive traits.
- This means, the more biodiversity, the higher stability for the world.
- More biodiversity also gives us more food and medicine and natural resources.

Resistance

As discussed in previous chapters, some pests get a mutation in their DNA that make them resistant to pesticides (insect-killing sprays). While all his pest "friends" will get killed out by a pesticide, he will stay alive because of his mutation! Lucky him!

Common Ancestry

Evolutionists say that similar features in different animals may point to a common ancestor. They think that different species evolved from this common ancestor.

Here are types of similarities scientists look for to determine common ancestors. This list is listed from most to least reliable.

1. Similar DNA patterns
   • Tested with **GEL ELECTROPHORESIS**
2. Similar protein and amino acid patterns
3. Similar chemical makeup
   • Tested with **INDICATOR TESTS**
4. Similar colors
   • Tested with **PAPER CHROMATOGRAPHY**
5. Similar structure and physical characteristics

We'll learn more about these things in the next unit.

Common Ancestry — Bone Structure

Evolutionists say that similar bone structure in different animals may point to a common ancestor.

In their studies, they use fossils.

- **FOSSILS:** Remains of dead organisms

CAT, WHALE, BAT

- **Purpose:** Used to see what living things were long ago compared to the same species today.

 Skeletons

Relatives

- There are more DNA similarities in relatives than strangers.

- Compared to all other relatives, siblings have the most similar DNA since they each inherit half the genes from the father and half from the mother.

- A child gets half the genes of each parent. So the DNA of a parent and child will be similar, but not as similar as a sibling.

- The closer the relative, the more similarities.

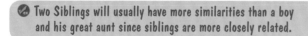

Two Siblings will usually have more similarities than a boy and his great aunt since siblings are more closely related.

Relatives Diagram

- The Relatives Diagram looks like a family tree

- Bottom (letter A) is the oldest generation/ ancestor

- Closer relatives have more similar genes/DNA
  Therefore: 1) **D&E** will have the most similar genes.
              2) **F&G** will have the most similar genes.
              3) **B&C** will have the most similar genes.

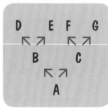

- This is a common Regents question. They'll ask which relatives' DNA are the most similar.

Relatives Diagram

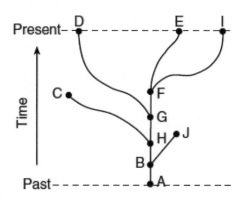

- A is the oldest ancestor in this tree

- D, E, and I are the youngest organisms on this tree

- C and J became extinct. Their chain stopped abruptly. They probably did not adapt to the environment

- E and I have the most similar DNA

- F is E and I's parent, so he will have similar genes to them

Congratulations!
You just finished most of the material.
All you have left is LABS!

Hurray!

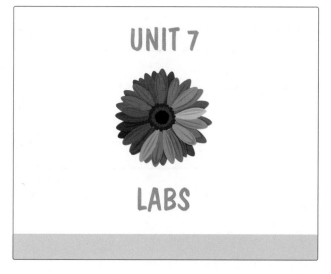

Every regents has Part D, a full section on labs. These labs are done in class and you are expected to know information based on the experiments. This unit will give you that info. Good luck!

PART 1:
Lab Skills

This unit is split into 2 sections:

1. General lab skills
2. Actual labs

Let's start with the general labs skills. Have fun!

Scientific Method

SCIENTIFIC METHOD: Method for performing scientific investigations

<u>Scientific Method steps:</u>

1. **OBSERVATION:** Use your five senses or tools to observe nature

We'll explore each step separately.

2. **HYPOTHESIS/INFERENCE:** Conclusion based on observation

3. **INDEPENDENT & DEPENDENT VARIABLE:** You change one variable (independent variable) and then measure how it affected the other variable (dependent variable).

Scientific Method

4. EXPERIMENTAL & CONTROL GROUPS: Set up one group that gets tested with the independent variable and another group that doesn't get tested with the independent variable

5. Repeat the experiment many, many times on many subjects

6. GATHER DATA: Gather the results of the experiment and analyze them

7. Create a **CONCLUSION:** A statement that describes the effect of the independent variable on the dependent variable based on the data

8. PEER REVIEW: Get your work checked over by a peer

Observation

- Observe nature. Look at different things around you using your 5 senses or by using tools, such as a microscope, thermometer, or ruler

- Gather all information that you can

- You can also gather information from other people's past experiments.

> ✏️ You see plants growing. You measure the plant growth with a ruler. You see people watering plants.

Hypothesis

- **HYPOTHESIS:** What you think will happen based on your observation and scientific knowledge
- Format of hypothesis: Action affect results

  > If you water a plant (action), it will grow (result).

- <u>Do **NOT**</u> write your hypothesis in question format. Do not write "I think". Write it as fact even though it's only a guess. It's OK if you'll be wrong.
- Must include the **INDEPENDENT** and **DEPENDENT** variables in the hypothesis

Independent & Dependent Variables

- **INDEPENDENT VARIABLE:** Variable that is changed in the experiment

   Watering the plant

  Only ONE independent variable can be tested in each experiment. Otherwise, the results will not be accurate.

- **DEPENDENT VARIABLE:** Variable that is affected by the experiment

   Growth of the plant

  The <u>d</u>ata will be measuring the <u>d</u>ependent variable.

Testing One Variable

You can only test **ONE** variable in every experiement. All other variables must remain the same!

> ✅ If you are testing the effect of watering a plant, all other variables must stay the same: It must be the same plant, same temperature, same soil etc.

> ❌ If you are testing the effect of a medicine on a sick person, all other variables must stay the same: Same gender, same weight, same height, same age

If you test other variables, the results won't be accurate because they will be based on other factors.

Hypothesis & Independent / Dependent Variables

The hypothesis states the effect of the **INDEPENDENT VARIABLE** on the **DEPENDENT VARIABLE**.

> ✐ If you water a plant (independent), it will help it grow (dependent).

Independent variable → Dependent Variable

Tip for writing a valid hypothesis:

Write: If _____ (independent variable), then _____ (dependent variable).

> ✐ If you water a plant (independent), it will help it grow (dependent).

Experimental and Control Groups

Whenever you set up an experiment, you make 2 groups:

Group 1: EXPERIMENTAL GROUP — gets tested with the independent variable

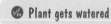

Plant gets watered

Group 2: CONTROL GROUP — does not get tested with the independent variable

- The purpose of this group is to be able to compare normal results to the experimental results.
- The control group often gets opposite treatment than the experimental group.

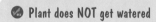

Plant does NOT get watered

Multiple Experimental Groups

Sometimes the regents asks you to set up an experiment with 3 experimental groups.

Yikes! Aren't we only allowed to test one variable in every experiment?

Yes, you still test one variable (the independent variable), just in different quantities

> 🦴 If your independent variable is watering a plant, your experimental groups can be:
> 1. Water the plant 1 Tablespoon a day
> 2. Water the plant 2 Tablespoons a day
> 3. Water the plant 3 Tablespoons a day
> (Control group: Plant gets NO water.)

> 🦴 If you're testing the affects of salt on fish growth, your groups can be
> 2 cups of salt, 4 cups of salt, 6 cups of salt (Control group gets no salt.)
> OR 2% salt solution, 5% salt solution, 10% salt solution (Control group gets no salt.)

Placebo

Let's say I want to try out a vitamin that claims to make you smarter.
I set up an experimental group who gets the vitamin and a control group who doesn't get the vitamin.
The experimental group may score higher on their tests because of the psychological effects!

In order to solve this issue, whenever we experiment with medicine, vitamins, or creams, we give the control group a **PLACEBO**: a fake pill/fake cream. It looks the same as the real medicine or cream, but it's really fake. The people we are experimenting on are not told who got the real medicine or cream and who got the placebo. This way, we know the results are accurate and not psychological.

 Experimental group: real medicine, Control group: placebo

Repeat Trials on Many Subjects

If we would test an experiment one time on one person or one plant, the results wouldn't be fair. Maybe the results were just a fluke.

In order to make sure you get accurate results, you must perform the experiment many times (many trials) on many subjects.

> Have 400 people take a medicine. Then repeat the experiment 400 times.

Data

- Now that you did all the steps of the experiment, you need to measure the results.
- DATA: facts and statistics collected together
- The data will always be in a measureable/quantitative form

  > ✏ Growth of plant in inches, marks on test, pulse rate

- The <u>d</u>ata will always be measuring the <u>d</u>ependent variable.

  > ✏ In an experiment testing the affect on how watering a plant effects its growth, the data will be: growth of plant

Organizing Data

Information can be organized into:

- Data Tables
- Diagrams
- Charts
- Graphs

Change in Length

Concentration of Sugar Solution (grams per liter)	Original Length of Potato Pieces (mm)	Average Length After 24 Hours (mm)
0	50.0	52.0
5	50.0	44.0
8	50.0	43.5
10	50.0	42.5

Data table: Organizes information
from an experiment

- Sometimes the regents will give you data and ask you to put the data in order in a chart.

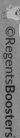

Types of Graphs

Graphs organize the data (from a data table) to show relationships
There are different types of graphs:

BAR GRAPH

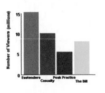

PIE GRAPH

LINE GRAPH

HISTOGRAM

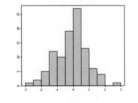

Line Graph Rules

- Dependent variable is plotted on the Y Axis (vertical).
- Independent variable is plotted on the X Axis (horizontal).
- The intervals in between each number on each axis must be equal.

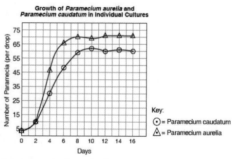

- Title and label the axis, including the unit of measurement (Example: cm. or inches etc.),
- Plot points correctly, then connect the points
- Sometimes the question asks you to graph two different things, represented by 2 different shapes on one graph. Graph the first variable using the correct symbol and attach the points. Then go on to the next variable using the other symbol and connect those dots. Do not attach both variable's points.

©RegentsBoosters™

Line Graph Rules

- When the regents asks you to place a scale on the graph, that means to label/number the axes
- Line graphs show relationships between different factors
- When graphing 2 groups from a data table:
1. First figure out the coordinates for the first group:
   In this chart: (0,17), (1,18), (2,17), (5,18), and (10,16)
2. Then find the next set of coordinates for the next group:
   In this chart: (0,18), (1,15), (2,11), (5,4), and (10,1)

Growth of Bacterial Colonies

Petri Dish	Exposure Time to UV Light	Number of Bacterial Colonies on Screened Side	Number of Bacterial Colonies on Unscreened Side
1	No Exposure (0.0 minutes)	17	18
2	1.0 minutes	18	15
3	2.0 minutes	17	11
4	5.0 minutes	18	4
5	10.0 minutes	16	1

Almost every regents has a graph question. Practice graph questions!

Bar Graphs vs. Histograph

Salaries Per Hour

BAR GRAPH:
Bar graphs are drawn in bar form. Well yeh, that makes sense!
Label the axis with consistent intervals. Draw the bars as seen in the chart. (Sometimes one axis will have titles instead of numbers.)

Salaries Per Hour

HISTOGRAPH:
Same as a bar graph, except there's an interval of numbers (See x axis in this graph.)

Conclusion

- After you analyze the data, you write your conclusion.
- You can NEVER change your data to match your hypothesis.
- Your conclusion includes the independent and dependent variable including the results that occurred
- Your conclusion is based on your DATA.

> ✅ If you water a plant, it will grow taller.

- If your hypothesis was correct, your conclusion supports your hypothesis. You can use your hypothesis as a conclusion.
- If your hypothesis was incorrect, your conclusion does NOT support your hypothesis.
- If your hypothesis is incorrect and you want to make your experiment correct, you will need to start a new experiment and test a new hypothesis and follow the rest of the steps of the scientific method. You CANNOT change your hypothesis after you complete your experiment.

Peer Review

After you finish your experiment, you show it to peers for feedback

Why?

- To make sure you didn't make any mistakes

- To make sure you weren't prejudiced with your results

Theory

- After an experiment is conducted many, many times and there is strong evidence that the conclusion is correct, it can be called a **THEORY**.

- A theory is not a fact. Therefore, as more scientific data becomes available, theories can change over time.

Valid Experiment

In order for an experiment to be considered **VALID** it must:

- Contain a control group
- Test only **ONE** variable
- Repeat the experiment many times on many subjects
- You cannot change the data to match your hypothesis

A valid experiment is sometimes called a **CONTROLLED EXPERIMENT**
(Don't confuse with the control group.)

If any of these factors are missing, the experiment is **INVALID**

Example of a Scientific Experiment

An experiment is done to see if eating breakfast increases brain ability.

- **HYPOTHESIS:** People who eat breakfast will score higher marks on their tests.

- **EXPERIMENTAL GROUP:** Group of 100 14 year olds who ate breakfast before taking the living environment regents

- **CONTROL GROUP:** Group of 100 14 year olds who fasted before taking the living environment regents

- All the variables in the experimental group and control group are constant (age, test, health).

- **INDEPENDENT VARIABLE:** Students eating breakfast or not

- **DEPENDENT VARIABLE:** Marks on the tests

Example of a Scientific Experiment
(Continued)

- **CONCLUSION:**

  - If the students who ate breakfast did significantly better than those who didn't:
    Eating breakfast increases brain ability.

  - If the students who ate breakfast didn't do better than those who fasted:
    Eating breakfast doesn't increase brain ability.

Safety in the Lab

- Don't mix chemicals on your own.

- Long hair should be in a ponytail when working with an open flame.

- Safety goggles should be worn while experimenting with heat and specific chemicals.

- Wash your hands after dissections.

- Don't heat test tubes with their stoppers on. They could burst.

Measurements: Length

- **METRIC RULER:** Measures length

- **Units of measurements:** cm (centimeters) or mm (millimeters)

> Conversion of mm to cm: 10mm = 1cm

 5 cm = 50mm
20mm = 2cm

- **Micrometers (μm):** tiny units to measure objects through a microscope

> Conversion of mm to μm: 1mm = 1000 μm

Measurements: Volume

- **GRADUATED CYLINDER:** Measures liquid's volume/the space it takes up

- Liquid in the graduated cylinder has a curved surface. Measure from the bottom of the curved line.

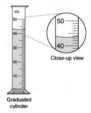

- Units of measurements: mL (milliliters)

  (Volume can also be measured in Liters)

Measurements: Temperature

In lab, Celsius is usually used.

- Freezing point of water: 0° C
- Boiling point of water: 100° C
- Normal Human body: 37° C

Measurements: Mass

- **MASS:** Amount of matter in something
- **BALANCE:** Tool that measures mass
- <u>To Find the Mass of a Substance:</u>

ELECTRIC BALANCE

  1. Weigh the dish you are going to put the substance on.
  2. Weigh the substance <u>together</u> with the dish.
  3. Subtract the weight of the dish from the total you got in #2.

TRIPLE BEAM BALANCE

- If you're using weighing paper, use the same instructions as if using a dish.

Microscopes

MICROSCOPE: Uses lens(es) to make objects easier to see

MAGNIFICATION: Ability of a microscope to make an object look bigger

Compound Microscopes

COMPOUND MICROSCOPE: Light microscope with two lenses

- Lenses: Ocular, Objective
- Light Source: A mirror or light bulb
  Light passes through the specimen ⇨ through the objective lens
  ⇨ through the eyepiece that you can see the object
- The image is the picture you see.

> To find total magnification of an object:
> Magnification of Ocular lens X Magnification of Objective lens

> ✅ If the eyepiece (ocular lens) = 10x magnification and the objective lens = 40x
> magnification
> Total Magnification: 10 X 40 = 400. 400 would be the answer.

Compound Microscope Parts

Memorize these parts of the microscope and how they look on the diagram:

- **OCULAR LENS:** eyepiece

- **OBJECTIVE LENSES:** makes images bigger or smaller

- **DIAPHRAGM:** makes the image brighter or less bright

- **COURSE and FINE ADJUSTMENT KNOBS:** used to focus the image/ make it more clear and detailed

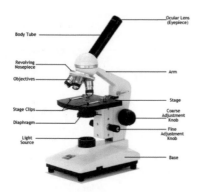

Low Power Objective Lens

When using a compund microscope, you can either use the HIGH POWER objective lens or the LOW POWER objective lens.

The LOW POWER OBJECTIVE LENS is:
- <u>L</u>ow power objective lens is the shortest lens (low=short)
- <u>L</u>ighter (brighter) than the high power lens
- A <u>l</u>ot of the specimen (is visible) — it's great for an overview of an object.

💡 HINT: LLL: <u>l</u>ow power lens, <u>l</u>ighter, a <u>l</u>ot of the speciment is visible

How to use the microscope under low power objective lens:
1. Turn the coarse adjustment knob to focus the image
   - Coarse adjustment can only be used under low power
2. After, adjust the fine adjustment knob to focus image more
   - Fine adjustment: Could be used with high power/ low power
   - Center the image so that you can see entire specimen
   - When you switch from high to low power, the lens zooms into the center.

💡 HINT: People are sometimes coarse with "low power" people :(

High Power Objective Lens

The **HIGH POWER OBJECTIVE LENS** shows a high magnification of the specimin.

- It has the longest lens (high=long).
- Part of specimen will be larger and more detailed.
- Less of the whole specimen is visible (because we're zoomed in to a specific part).
- The specimin appears darker than the low power lens. (You can adjust the diaphragm to add light.)

When using the high power objective lens, only use the fine adjustment knob.

- The course adjustment knob <u>cannot</u> be used in high power. Using the coarse adjustment can break the microscope! Since the high objective lens is longer, if you adjust the course adjustment knob, it will hit the glass and break it! :(

⬤ This is a common regents question.

🔵 HINT: The high power objective is like the zoom feature on a camera. It focuses clearly a smaller area.

🔵 HINT: With <u>high power</u> people, use only <u>fine</u> materials.

Low Power Vs. High Power
(Compound Microscope)

<u>LOW POWER</u>

- <u>L</u>ow lens (short lens)
- <u>Li</u>ghter (brighter)
- A <u>L</u>ot of specimen (is visible)

<u>HIGH POWER</u>

- Less of the specimen is visible
- The "zoomed in area" is clear and large
- Darker (You can adjust the diaphragm to add light.)

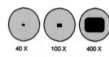

40 X   100 X   400 X

Different magnifications of the same specimin

- The lower power shows the whole specimin.
- The higher power shows less of the specimin, but a bigger and more detailed view of a part of the specimin.

Techniques for Using Compound Microscopes

- Image will be upside down and backwards

- To move image: Move slide in the opposite direction.

  > To move image up, pull slide down.
  > To move image to the right, pull slide to the left.

- As you increase the magnification, the field gets darker. Increase the amount of light (when going from low power to high power).

- Always start with low power (even if you'll eventually need high power).

Wet Mount Slides

1. Add a small amount of water to the center of a clean, glass slide using a pipette (eye dropper).

2. Put the object into the water.

3. Use forceps to slowly put a cover slip on the object at a 45-degree angle. This is done to reduce air bubbles.

   ❗ This is a common regents question.

Staining Specimens

Purpose:
Stains make the cell structures more visible when using a microscope

Common Stains:
1. <u>Iodine</u>: Darkens certain cell structures.
2. <u>Methylene blue</u>: Stains structures in the cell

To Add a Stain (like methylene blue or salt):
1. Drip a drop of stain beside one edge of the cover slip
2. Place a small piece of paper towel on the opposite side of the cover slip (the paper towel will pull the stain across the slide).

I better make sure not to get stained.

Gel Electrophoresis

GEL ELECTROPHORESIS: The process used to separate mixtures of large molecules from the base sequence according to size using an electric current

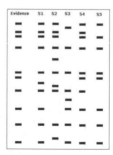

Electrophoresis results

Gel Electrophoresis Process

1. Special enzymes are used to cut the DNA at specific points in the base ⇨ pieces are different lengths and sizes.

   ✅ Between every CC and GG, you make a cut

2. DNA samples are put into the wells of a semisolid gel In the first well, you put in the known substance.

   ✅ Blood found at a crime scene

   In the other wells, you put the unknown substances.

   ✅ Blood of 3 suspects (Tim, Dick, Harry)

Gel Electrophoresis
Process & Results

3. Turn on the electric current of the machine ⇨ all DNA becomes negatively charged.
   The bottoms of the wells are positively charged ⇨ pieces are attracted to the bottom. The smallest pieces move the furthest and fastest.

🔵 Big, fat, "klutzy" pieces have a hard time swimming to the bottom ⇨ they'll stay closer to the top.

4. When the process is finished, you get a printout of each well. Compare wells #2,3,4(unknown) to well #1 (known). Whichever matches to well #1, you know that substance is the same substance as well #1.

✏️ Harry did the crime because his blood matches exactly to the blood found at the crime scene

WELL 1 Blood found at crime scene	WELL 2 Tom	WELL 3 Dick	WELL 4 Harry
▬ ▬ ▬	▬ ▬	▬ ▬	▬ ▬ ▬

Purpose of DNA Gel Electrophoresis

- Prove PATERNITY (who the father of a child is) or MATERNITY (who the mother of a child is)

- For criminal investigations: who's guilty, who isn't

- Identify which genes are responsible for a specific genetic disease. Maybe they could find a cure for it.

Paper Chromatography

PURPOSE:
Separates mixtures of molecules

The mixture is usually a solution of liquid plant pigments with different kinds of chlorophylls and other pigments.

Paper Chromatography Procedure

1. Put a small sample of mixture on the chromatography paper on top of the line of the solvent mixture.

2. The solvent moves through the paper and dissolves the mixture. The colors will get separated into a pattern.

   • Some parts of the solvent move a lesser distance.

   • Some parts of the solvent move a greater distance.

   • If the water level was above the pigment spots, the pigment will wash away and not go up the paper.

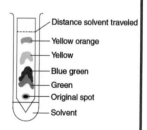

Distance solvent traveled
Yellow orange
Yellow
Blue green
Green
Original spot
Solvent

Paper Chromatography Results

- The specific mixture put on chromatography paper will always separate into the same pattern at the same rate.

  Therefore, we can use chromatography to identify an unknown substance by using the results of this test (the rate it separated, and the colors it separated into)

  ✅ Unknown substance 2 must be tulip juice.

🔵 Paper Chromatography is not as accurate as gel electrophoresis.

TULIP JUICE	UNKNOWN 1	UNKNOWN 2
yellow	blue	yellow
red	orange	red
green	green	green

Indicators

INDICATORS: Substances that change colors when they come into contact with certain chemicals

TYPES:

1. **PH PAPER:** Dip the PH paper into a solution. The color of the paper is then matched to a color scale. Different colors indicate whether the substance is basic or acidic.
2. **IODINE (LUGOL'S) SOLUTION:** If the color changes from golden brown to bluish blackish, it means there's starch in the solution.
- Other indicators show if there's sugar or carbon dioxide in a solution.

The swimming pool maintenance crew uses indicators to tell them if there's chlorine in the water.

PH PAPER

Dichotomous Key

DICHOTOMOUS KEY: Sequence of steps that helps you identify or categorize a living thing

The key will have choices that lead the user to the correct name of the given item. Follow the steps as you go along. Do not skip steps — or your answer will not be correct.

Practice examples!

Sample Dichotomous Key

1a. If it has a door	....	it's a house
1b. If it has no door	....	go to 2
2a. If it has a trunk	....	it's a tree
2b. If it has no trunk	....	go to 3
3a. If it has rays	....	it's a sun
3b. If it has no rays	....	it's a flower

Dichotomous Key Question

1. The dichotomous key shown below can be used to identify birds *W*, *X*, *Y*, and *Z*.

Bird W   Bird X   Bird Y   Bird Z

Dichotomous Key to Representative Birds
1. a. The beak is relatively long and slender............................*Certhidea* b. The beak is relatively stout and heavy..................................go to 2
2. a. The bottom surface of the lower beak is flat and straight .....*Geospiza* b. The bottom surface of the lower beak is curved ....go to 3
3. a. The lower edge of the upper beak has a distinct bend ....*Camarhynchus* b. The lower edge of the upper beak is mostly flat ....*Platyspiza*

Bird *X most likely* falls under which classification?

A) *Certhidea*   B) *Geospiza*   C) *Camarhynchus* D) *Platyspiza*

Answer: D

Making Your Own Dichotomous Key

You will be given several items.

In the examples below, the items are: House, tree, sun, flower.

1. Choose 1 item and 1 characteristic of that item.

   Write: 1a. If it has characteristic "X", it is "item 1"

   ✅ 1a. If it has a door.............................it's a house

2. Write: If it doesn't have "characteristic X", go to 2

   ✅ 1b. If it has no door.................................go to 2

3. Choose your 2nd item and 1 characteristic of that item

   Write: 2a. If it has characteristic "Y", it is "item 2"

   ✅ 2a. If it has a trunk..............................it's a tree

Making Your Own Dichotomous Key (Continued)

4. Write: If it doesn't have "characteristic Y", go to 3

   ✏️ 2b. If it has no trunk.......................go to 3

5. Choose your 3rd item and 1 characteristic of that item

   3a. If it has characteristic "Q", it is "item 3"

   ✏️ 3a. If it has rays...............................it's a sun

6. Write: If it doesn't have "characteristic Q", it is "item 4"

   ✏️ 3b. If it has no rays.......................it's a flower

⬤ Make sure the characteristic you choose only fits 1 item.

   ✏️ Don't say "If it has leaves..." because that could be a tree or a flower.

Making Your Own Dichotomous Key Example

Species E   Species F

Dichotomous Key

1. a. has small wings .... go to 2
   b. has large wings .... go to 3

2. a. has a single pair of wings .... Species A
   b. has a double pair of wings .... Species B

3. a. has a double pair of wings .... go to 4
   b. has a single pair of wings .... Species C

4. a. has spots .... go to 5
   b. does not have spots .... Species D

5. a. _____ .... Species E

   b. _____ .... Species F

A dichotomous key to these six species is shown above. Complete the missing 5.a. and 5.b. so that the key is complete for all six species.

Answer:

5. a. has white or clear or light wings

5. b. has shaded or black or dark wings

Dissection

DISSECTION: Examining preserved specimens

<u>Purpose:</u>
- Identify similarities & differences amongst species
- Identify internal structures of organisms

<u>Tools:</u>
1. **TEASING/DISSECTING NEEDLE:** Pulls muscle tissue apart
2. **DISSECTING SCISSORS:** Cuts the tissue
3. **SCALPEL:** Knife used to cut/slice tissue

Part 2
Actual Labs

Till now, we learned the lab skills. Now let's discuss how they apply to each of the 4 NYS labs.

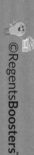

Lab #1
Relationships & Biodiversity Overview

- Botana curus is a plant that was found to cure cancer. Since this plant grows very slowly, and is endangered, we are trying to find another plant that can cure cancer. Since plants that are related to B. Curus are more likely to cure cancer also, we are trying to find a plant that has structural and molecular similarities. In this lab, you test 3 plants in several ways to find which one is the most similar to B. Curus.

- B. Curus does not cure cancer. It is a made up plant.

- <u>Lab Procedure:</u>

1. First you just look at the plants and the seeds to see what structural characteristics are the most similar to B. Curus.

   Same color, same shape leaf, same shape plant.

Relationships & Biodiversity Overview
(Continued)

2. Look at the stems through a microscope. See if the arrangement of the bundles is scattered or circular.

3. Do Paper Chromatography for each plant (B. Curus and the 3 other plants) and note which colors from each plant show up on the paper.

4. Test each plant with the Indicator test for Enzyme M.

5. Gel electrophoresis test:

   You will be given strips of paper with DNA codes on it. You will be told to cut the paper in between given letters (Example: between CC&GG). Then, you place the strips of paper on a given chart indicating the size of the DNA fragment. You will find that 1 plant has the same pattern as B. Curus.

⬤ The letters don't have to match; only the sizes (amount of DNA bases) of the fragments have to.

Relationships & Biodiversity Overview
(Continued)

○ Know how to translate the DNA code to RNA, and to find the amino acid it produces on the "Universal Genetic Code Chart" that will be given to you.

• You will record all your results on a chart, comparing the 3 plants to B.Curus.

• Molecular evidence is stronger the structural evidence.

• The plants that look the most similar to B. Curus are not the ones that really are the most similar.

Here's a list going from the most to the least accurate method to find similarities in plants:

1. Gel Electrophoresis

2. Paper Chromatography

3. Indicator test

4. Observations with and without a microscope (not accurate at all)

Facts to Know on Relationships & Biodiversity

- If two species have similarities, they may be related
- The more similarities they show, the more closely they may be related
- Types of similarities
    1. Similar characteristics/structure – how they look
    2. Similar proteins/amino acid patterns
    3. Similar DNA patterns (gel electrophoresis)
    4. Chemical similarities – contains similar enzymes (Enzyme indicator tests)
    5. Similar colors (found through paper chromatography)
- Most reliable: similar DNA
- Least reliable: similar structure
- If one known species produces something, you can assume that a similar species will produce the same thing.
- Endangered species should be protected because they may offer benefits to humans.

Relationships & Biodiversity Skills

- Translating the DNA code (base pairing from DNA to RNA to get mRNA sequences-G,C & A,U —not T). Similar DNA may mean common ancestry.

- Branch Tree Diagrams to show relationships (See Unit 6.)

- Human actions affect biodiversity

- Benefits of biodiversity (We can get different benefits from the different organisms, like medicines and natural resources.)

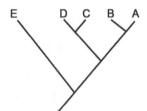

Questions 1

? **How did you use your lab data to find which plant species is the closest relative to B. Curus?**

Eyes, microscope, enzyme test, paper chromatography, gel electrophoresis

? **Why is finding a close relative to B. Curus important?**

B. Curis is dying out. We want another plant that has similar genes to cure cancer.

? **What are some forms of structural evidence that could be used to prove biological relationships?**

Similarities in flower form, leaf shape, color

Questions 2

❓ Why is structural evidence less dependable that molecular evidence when finding biological relationships?

Structural evidence may not always be objective, while molecular evidence is.

❓ Why do closely related species share similarities?

They may come from the same ancestor, and therefore have similar DNA.

❓ List ways biodiversity could be lost.

Deforestation, overhunting, pollution

❓ Why do tradeoffs have to be considered when destroying habitats of species?

Even though trees are very useful in day-to-day life (paper, furniture...), by cutting down trees we may be lessening biodiversity!

Lab #2
Making Connections Overview

You are finding patterns in pulse rates and muscle fatigue.

- ## LAB PROCEDURE:

1. Take your pulse rate by counting how many pulses you feel in 20 seconds. Then multiply this # by 3. Your answer is your pulse rate.

> ✅ I felt 22 pulses in 20 seconds⇨ 22 x 3 = 66. My pulse is 66.
> Repeat this step 3 times for accuracy.

2. Record the pulse rates of your class on a data table and histogram and see if you find any patterns. (Do the taller students have a higher pulse rate?...)

Making Connections Overview
(Continued)

3. Test your pulse rate after doing exercise. Your pulse will go up because your heart is beating faster, so your pulse rate is faster.

4. Count how many times you can squeeze a clothespin in 1 minute. Repeat this test again right after the first round.

- Were you able to squeeze more or less times in the second round?
  - If you squeezed the clothespin less times in your second round, it could mean that your muscles produced waste (lactic acid) that slowed down your performance. This is called MUSCLE FATIGUE.
  - If you squeezed the clothespin more times in your second time, it could mean that your muscles got more circulation from the first round and were able to pick up speed in the second round.

Skills

- Taking pulse rate

- Tables and graphs

- Designing a controlled experiment and forming conclusions

- Finding an average: add all the numbers together, then divide by the amount of numbers there are.

> Find the average of these numbers: 60, 90, 120, 110
>
> $60+90+120+110=380$
>
> $380/4=\underline{95}$
>
> We divided by 4 because there were 4 numbers.

Questions 1

? **What does your pulse rate tell you about what's going on in your body?**

How fast your heart is beating

? **Why would higher activity affect a person's pulse rate?**

Respiration must be done at a faster rate. The oxygen needed to do respiration must be brought to the cells faster ⇨ the heart must beat faster

? **What is muscle fatigue? Why does it happen? How can you stop it?**

During strenuous exercise, your body can't supply enough oxygen to the muscles fast enough. Since there isn't enough oxygen, the muscles produce lactic aid, which accumulates in the muscles, and slows down your muscle activity. To stop it, breathe a lot to increase your oxygen during exercising.

? **Why do different people have different resting pulse rates?**

Because people have different fitness levels, different weight

Questions 2

❓ **Why do different people get muscle fatigue from different amounts of activity?**

Because people have different fitness levels. By a fit person, the muscles build up less waste (lactic acid). Waste causes muscle fatigue. By a non fit person, the muscles build up much more waste.

❓ **How would you prove that a person's pulse rate goes up when watching an exciting game?**

Perform a controlled experiment, repeat a few times, and record your results.

❓ **How do you know if a company's claim about a product is true or not?**

By seeing results of scientific testing, you see in black and white if the claim is true or not — it's purely objective.

❗ **On the regents, they may ask you to write out how you would perform an experiment. Make sure you write that you repeated the experiment many times.**

Lab #3
The Beaks of Finches Overview

- Evolution occurred with finches on the Galapagos Islands so that species have different beak shapes and sizes

- Animals compete for limited resources

- Natural selection occurs and the adapted organisms survive

- There are different adaptations (beak size, beak type) for different niches (different foods, size of seed...)

- Different variation of finches help more finches survive:

   1. Beak size and shape

   2. Color (camouflage)

   3. Speed/strength

   4. Smaller finches need less food

   - Tools: forceps/tweezers/large hair clips mimic beaks of finches

The Beaks of Finches Overview
(Continued)

- What we did: Used different tools to represent different finch species competing for food.

- What we learned: Different environmental conditions (food) favored different species of finch, allowing some to survive and reproduce, but not others.

- Know how to read the Beaks of Finches Chart

Questions 1

❓ What procedures did you do in this lab that are associated with: variation, competition, survival, adaptation, environment in natural selection?

a. Variation: different beaks, different size seeds

b. Competition: more than one bird feeding at one bowl

c. Survival: each bird trying to get enough food to survive

d. Adaptation: particular characteristics of "beaks"

e. Environment: students, seeds, dishes are part of environment

Questions 2

① **How is being assigned a specific tool more like adaptation then choosing the tool?**

The finches don't get to choose which beak they are born with.

② **How could the characteristics of a certain bird's beak affect its survival?**

If that beak is able to eat the food on the island easily and quickly, it will survive (and vice versa).

③ **How could a beak be useful in one environment and useless in another environment?**

A tiny pointy beak would be great for an environment with tiny seeds, but useless in an environment with huge seeds that it can't pick up.

Questions 3

● How would migration to a new environment help an organism that wasn't able to survive in its old environment?

There may be less competition in the new environment.

● Why can't a bird with a beak that isn't well adapted just get a new better beak?

The DNA in an organism determines its features. A bird can't choose to change its DNA!

Questions 4

❓ **What are 2 other characteristics of a bird (besides for its beak) that could vary in individuals in a population that could help the bird survive?**

1. Rate of flying — If food is scarce, he can get to the food before other birds.
2. Camouflage ability — It will be harder for predators to find the bird and kill it.

❓ **Which 2 species from the beak chart would probably compete if there were limited food?**

Choose 2 birds that eat the same food (large tree finch and small tree finch).

⬤ The more competition, the lower the survival rate.

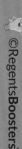

Lab #4
Diffusion Through a Membrane Overview

<u>Lab Procedure:</u> Create a model cell by tying the ends of a dialysis tube or plastic bag tightly with a string.

♦ Put glucose and starch in the inside of the tube (colorless) and water and starch indicator on the outside of the tube (light brown/amber).

♦ Predict which substances will move into the cell and which will move out of the cell.

This lab is designed to help you understand the process of diffusion.

Part 1: You make a model cell and test the selective permeability of the membrane.

Part 2: Then you add salt and distilled water to a red onion cell and observe the results.

Diffusion Through a Membrane Overview
(Continued)

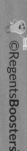

What we did:

- We made a model cell using dialysis tubing. The tubing represents the cell membrane since it's selective permeable (lets only some things in).

- Put glucose and starch inside your "cell"

- Put starch indicator (iodine) outside "cell". Starch indicator is amber colored. The starch indicator will tell us if there's starch present.

- We also tested each indicator with starch and glucose to see how each indicator reacts when combined with starch or glucose.

Lugol's iodine solution

dialysis tubing containing starch solution

water in beaker

Diffusion Through a Membrane Overview
(Continued)

Specific indicators tell us is a specifics substance is present. If the substance is present, the indicator changes colors.

STARCH INDICATOR SOLUTION	GLUCOSE INDICATOR SOLUTION
Also called Iodine Color: Amber Turns blue/black in presence of starch	Also called Benedict Solution Color: Blue Turns orange/green in presence of glucose <u>when heated</u>

- When starch is placed together with starch indicator, the starch indicator turns blue/black
- When glucose is placed together with glucose indicator solution and heated, the glucose indicator turns orange/green

Diffusion Through a Membrane Overview
(Continued)

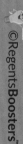

What happened?
- The inside of cell turned black because starch indicator diffused into the cell
- Starch did not diffuse out of the "cell" because the surrounding liquid did not change colors. It remained amber. It did not turn black. (Starch is too big to diffuse out.)
- Glucose did diffuse out of the membrane and therefore the outside of the cell turned colors in the presence of glucose indicator. (Glucose is small enough to diffuse out.)

Why?
- Small molecules (glucose) can diffuse through a membrane on their own
- Large molecule (starch) cannot diffuse through a membrane on their own (because they do not fit through).
- We use the indicators to identify the presence of specific substances. Indicators are small enough to diffuse into and out of the cell.

Diffusion Through a Membrane Overview
(Continued)

This is the second part of the lab:

- Place an onion on a slide

- Lower the coverslip carefully to avoid air bubbles

- Place a paper towel at one end of coverslip and drop a few drops of distilled (regular) <u>water</u> at the other end of the coverslip. The onion cell's cytoplasm and cell membrane will swell (grow) because water diffused into the onion cell.

- Now, place a paper towel at one end of coverslip and drop a few drops of <u>salt</u> water at the other end of the coverslip. The salt water will diffuse through the onion cell. You will see the cytoplasm and cell membrane shrivel/shrink because water from the onion cell is diffusing out of the onion cell.

- The cell wall never changes its shape (even when placed in water or salt solution).

Diffusion Through a Membrane Overview
(Continued)

RED ONION CELL DIAGRAMS

When an onion cell is placed in salt water, the cell membrane and cytoplasm shrink. When the onion cell is placed back into distilled water, the cell membrane and cytoplasm grow.

Cells placed in salt water:

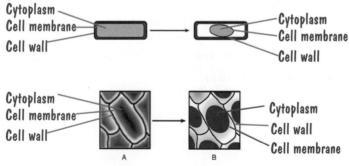

Questions 1

❓ **Why could you use a dialysis tube or plastic bag to substitute a cell membrane?**

They're both semi permeable: they let some things in and not others.

❓ **Give an example of an indicator. How could it identify if a specific substance is preset or not?**

Starch indicator solution: If the color changes from golden brown to bluish blackish, it means there's starch in the solution.

❓ **Why do you have to keep the onion skin material as flat as possible when making a wet mount slide?**

To prevent bubbling

Questions 2

? **Which direction do particles move in diffusion?**
High concentration to low concentration

? **How could the size of the molecule affect if it can/cannot pass through the membrane?**
If it's too big it won't be able to enter

? **How are cells affected by diffusion?**
Cell will have more or less of a molecule after diffusion happens.

? **What would happen if you put starch indicator solution inside the cell and filled the beaker with starch and glucose?**
The beaker would change to blue/black, but the inside of the cell would not change colors.

Questions 3

❓ What could be a negative affect of salt on the roads?

The salt can pull the water content out of plants and trees on the road and kill them, lowering biodiversity levels.

❓ If distilled water was given as I.V. instead of a salt solution, how could this disturb homeostasis?

The distilled water will diffuse into the person's cells and make them swell.

Questions 4

❓ Why are contractile vacuoles (a structure that pushes out extra water) needed in one-celled organisms living in fresh water, and not needed in one-celled organisms living in ocean water?

In fresh water: the water enters the organisms through diffusion and the contractile vacuoles need to push out the extra water.

In ocean water (salty water): the salt solution is either the same or less than inside this organism ⇨ there's no excess water entering the cell that needs to be pushed out.

❓ Give an example of diffusion in the body

Lungs: oxygen—diffuses into the lungs when inhaling

Carbon dioxide—diffuses out of the lungs when exhaling

Questions 5

? **How does salt affect the water content of the cell cytoplasm?**

When the cell is in a salty place, water diffuses out of the cell ⇨ cell shrinks.

? **How could distilled water be used to counteract the salt on living cells?**

Through osmosis, the water will go into the cell and return to normal size. If it stays in distilled water for too long, it will burst.

Hurray! You are finally finished!

Now make sure to do a lot of Regents
questions to test your knowledge and get
used to the questions.

For orders and comments, please email us at info@regentsboosters.com
or visit us at www.regentsboosters.com.
Check out our online Living Environment videos!